W9-BBV-435

This book is a concise introduction to the mathematical aspects of the origin, structure and evolution of the universe. The book begins with a brief overview of observational cosmology and general relativity, and goes on to discuss Friedmann models, the Hubble constant, models with a cosmological constant, singularities, the early universe, inflation and quantum cosmology. This book is rounded off with a chapter on the distant future of the universe.

The book is written as a textbook for advanced undergraduates and beginning graduate students. It will also be of interest to cosmologists, astrophysicists, astronomers, applied mathematicians and mathematical physicists.

An introduction to
mathematical cosmology

An introduction to mathematical cosmology

J. N. ISLAM

Research Centre for Mathematical and Physical Sciences,
University of Chittagong, Bangladesh

CAMBRIDGE
UNIVERSITY PRESS

Published by the Press Syndicate of the University of Cambridge
The Pitt Building, Trumpington Street, Cambridge CB2 1RP
40 West 20th Street, New York NY 10011, USA
10 Stamford Road, Oakleigh, Melbourne 3166, Australia

© Cambridge University Press 1992

First published 1992
Reprinted 1993

Printed in the United Kingdom by Bell and Bain Ltd., Glasgow

A catalogue record for this book is available from the British Library

Library of Congress cataloguing in publication data
Islam, Jamal N.
An introduction to mathematical cosmology / J. N. Islam.
   p.   cm.
Includes bibliographical references (p.   ) and index.
ISBN 0-521-37385-9 (hardback). — ISBN 0-521-37760-9 (paper)
1. Cosmology—Mathematics.   I. Title.
QB891.I73   1991
520′.15′1—dc20   91-28090   CIP

ISBN 0 521   37385 9   hardback
ISBN 0 521   37760 9   paperback

KW

Contents

Preface

Ever since I wrote my semi-popular book *The Ultimate Fate of the Universe* I have been meaning to write a technical version of it. There are of course many good books on cosmology and it seemed doubtful to me whether the inclusion of a chapter on the distant future of the universe would itself justify another book. However, in recent years there have been two interesting developments in cosmology, namely inflationary models and quantum cosmology, with their connection with particle physics and quantum mechanics, and I believe the time is ripe for a book containing these topics. Accordingly, this book has a chapter each on inflationary models, quantum cosmology and the distant future of the universe (as well as a chapter on singularities not usually contained in the standard texts).

This is essentially an introductory book. None of the topics dealt with have been treated exhaustively. However, I have tried to include enough introductory material and references so that the reader can pursue the topic of his interest further.

A knowledge of general relativity is helpful; I have included a brief exposition of it in Chapter 2 for those who are not familiar with it. This material is very standard; the form given here is taken essentially from my book *Rotating Fields in General Relativity*.

In the process of writing this book, I discovered two exact cosmological solutions, one connecting radiation and matter dominated eras and the other representing an inflationary model for a sixth degree potential. These have been included in Sections 3.5 and 8.4 respectively as I believe they are new and have some physical relevance.

I am grateful to J. V. Narlikar and M. J. Rees for providing some useful references. I am indebted to a Cambridge University Press reader for helpful comments; the portion on observational cosmology has I believe

xi

improved considerably as a result of these comments. I am grateful to F. J. Dyson for his ideas included in the penultimate chapter. I thank Maureen Storey of Cambridge University Press for her efficient and constructive subediting.

I am grateful to my wife Suraiya and daughters Nargis and Sadaf and my son-in-law Kamel for support and encouragement during the period this book was written. I have discussed plans for my books with Mrs Mary Wraith, who kindly typed the manuscript for my first book. For more than three decades she has been friend, philosopher and mentor for me and my wife and in recent years a very affectionate godmother ('Goddy') to my daughters. This book is fondly dedicated to this remarkable person.

Jamal Nazrul Islam

Chittagong, 1991

1

Some basic concepts and an overview of cosmology

In this chapter we present an elementary discussion of some basic concepts in cosmology. Although the mathematical formalism is essential, some of the main ideas underlying the formalism are simple and it helps to have an intuitive and qualitative notion of these ideas.

Cosmology is the study of the large-scale structure and behaviour of the universe, that is, of the universe taken as a whole. The term 'as a whole' applied to the universe needs a precise definition, which will emerge in the course of this book. It will be sufficient for the present to note that one of the points that has emerged from cosmological studies in the last few decades is that the universe is not simply a random collection of irregularly distributed matter, but it is a single entity, all parts of which are in some sense in unison with all other parts. This, at any rate, is the view taken in the 'standard models' which will be our main concern. We may have to modify these assertions when considering the inflationary models in a later chapter.

When considering the large-scale structure of the universe, the basic constituents can be taken to be galaxies, which are congregations of about 10^{11} stars bound together by their mutual gravitational attraction. Galaxies tend to occur in groups called clusters, each cluster containing anything from a few to a few thousand galaxies. There is some evidence for the existence of clusters of clusters, but not much evidence of clusters of clusters of clusters or higher hierarchies. 'Superclusters' and voids (empty regions) have received much attention (see Chapter 4). Observations indicate that on the average galaxies are spread uniformly throughout the universe at any given time. This means that if we consider a portion of the universe which is large compared to the distance between typical nearest galaxies (this is of the order of a million light years), then the number of galaxies in that portion is roughly the same as the number in another portion with the same volume at any given time. This proviso 'at any given time' about the uniform distribution of galaxies is important because, as we shall see, the universe is in a dynamic state and so the number of galaxies in any given volume changes with time. The distribution of galaxies also appears to be isotropic about us, that is, it is the same, on the average, in all directions from us. If we make the assumption that we do not occupy a special position amongst the galaxies, we conclude that the distribution of galaxies is isotropic about any galaxy. It can be shown that if the distribution of galaxies is isotropic about every galaxy, then it is necessarily true that galaxies are spread uniformly throughout the universe.

We adopt here a working definition of the universe as the totality of galaxies causally connected to the galaxies that we observe. We assume that observers in the furthest-known galaxies would see distributions of galaxies around them similar to ours, and the furthest galaxies in their field of vision in the opposite direction to us would have similar distributions of galaxies around them, and so on. The totality of galaxies connected in this manner could be defined to be the universe.

E. P. Hubble discovered around 1930 (see, for example, Hubble (1929, 1936)) that the distant galaxies are moving away from us. The velocity of recession follows Hubble's law, according to which the velocity is proportional to distance. This rule is approximate because it does not hold for galaxies which are very near nor for those which are very far, for the following reasons. In addition to the systematic motion of recession every galaxy has a component of random motion. For nearby galaxies this random motion may be comparable to the systematic motion of recession and so nearby galaxies do not obey Hubble's law. The very distant galaxies also show departures from Hubble's law partly because light from the very distant galaxies was emitted billions of years ago and the systematic motion of galaxies in those epochs may have been significantly different from that of the present epoch. In fact by studying the departure from Hubble's law of the very distant galaxies one can get useful information about the overall structure and evolution of the universe, as we shall see.

Hubble discovered the velocity of recession of distant galaxies by studying their red-shifts, which will be described quantitatively later. The red-shift can be caused by other processes than the velocity of recession of the source. For example, if light is emitted by a source in a strong gravitational field and received by an observer in a weak gravitational field, the observer will see a red-shift. However, it seems unlikely that the red-shift of distant galaxies is gravitational in origin; for one thing these red-shifts are rather large for them to be gravitational and, secondly, it is difficult to understand the systematic increase with faintness on the basis of a gravitational origin. Thus the present consensus is that the red-shift is due to velocity of recession, but an alternative explanation of at least a part of these red-shifts on the basis of either gravitation or some hitherto unknown physical process cannot be completely ruled out.

The universe, as we have seen, appears to be homogeneous and isotropic as far as we can detect. These properties lead us to make an assumption about the model universe that we shall be studying, called the Cosmological Principle. According to this principle the universe is homogeneous everywhere and isotropic about every point in it. This is really an extrapolation from observation. This assumption is very important, and it is remarkable that the universe seems to obey it. This principle asserts what we have mentioned before, that the universe is not a random collection of galaxies, but it is a single entity.

The Cosmological Principle simplifies considerably the study of the large-scale structure of the universe. It implies, amongst other things, that the distance between any two typical galaxies has a universal factor, the same for any pair of galaxies (we will derive this in detail later). Consider any two galaxies A and B which are taking part in the general motion of expansion of the universe. The distance between these galaxies can be written as $f_{AB}R$, where f_{AB} is independent of time and R is a function of time. The constant f_{AB} depends on the galaxies A and B. Similarly, the distance between galaxies C and D is $f_{CD}R$, where the constant f_{CD} depends on the galaxies C and D. Thus if the distance between A and B changes by a certain factor in a definite period of time then the distance between C and D also changes by the same factor in that period of time. The large-scale structure and behaviour of the universe can be described by the single function R of time. One of the major current problems of cosmology is to determine the exact form of $R(t)$. The function $R(t)$ is called the scale factor or the radius of the universe. The latter term is somewhat misleading because, as we shall see, the universe may be infinite in its spatial extent in which case it will not have a finite radius. However, in some models the universe has finite spatial extent, in which case R is related to the maximum distance between two points in the universe.

It is helpful to consider the analogy of a spherical balloon which is expanding and which is uniformly covered on its surface with dots. The dots can be considered to correspond to 'galaxies' in a two-dimensional universe. As the balloon expands, all dots move away from each other and from any given dot all dots appear to move away with speeds which at any given time are proportional to the distance (along the surface). Let the radius of the balloon at time t be denoted by $R'(t)$. Consider two dots which subtend an angle θ_{AB} at the centre, the dots being denoted by A

and B (Fig. 1.1). The distance d_{AB} between the dots on a great circle is given by

$$d_{AB} = \theta_{AB} R'(t). \tag{1.1}$$

The speed v_{AB} with which A and B are moving relative to each other is given by

$$v_{AB} = \dot{d}_{AB} = \theta_{AB}\dot{R}' = d_{AB}(\dot{R}'/R'), \quad \dot{R}' \equiv \frac{dR'}{dt}, \text{ etc.} \tag{1.2}$$

Thus the relative speed of A and B around a great circle is proportional to the distance around the great circle, the factor of proportionality being $\dot{R}'/R'$, which is the same for any pair of dots. The distance around a great circle between any pair of dots has the same form, for example, $\theta_{CD}R'$, where θ_{CD} is the angle subtended at the centre by dots C and D. Because the expansion of the balloon is uniform, the angles θ_{AB}, θ_{CD}, etc. remain the same for all t. We thus have a close analogy between the model of an expanding universe and the expansion of a uniformly dotted spherical balloon. In the case of galaxies Hubble's law is approximate but for dots on a balloon the corresponding relation is strictly true. From (1.1) it follows that if the distance between A and B changes by a certain factor in any period of time, the distance between *any* pair of dots changes by the same factor in that period of time.

From the rate at which galaxies are receding from each other, it can be deduced that *all* galaxies must have been very close to each other *at the same time* in the past. Considering again the analogy of the balloon, it is like saying that the balloon must have started with zero radius and at this initial time all dots must have been on top of each other. For the universe

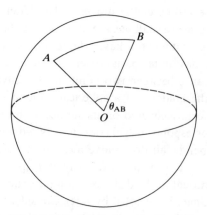

Fig. 1.1. Diagram to illustrate Equation (1.1).

it is believed that at this initial moment (some time between 10 and 20 billion years ago) there was a universal explosion, at every point of the universe, in which matter was thrown asunder violently. This was the 'big bang'. The explosion could have been at every point of an infinite or a finite universe. In the latter case the universe would have started from zero volume. An infinite universe remains infinite in spatial extent all the time down to the initial moment; as in the case of the finite universe, the matter becomes more and more dense and hot as one traces the history of the universe to the initial moment, which is a 'space-time singularity' about which we will learn more later. The universe is expanding now because of the initial explosion. There is not necessarily any force propelling the galaxies apart, but their motion can be explained as a remnant of the initial impetus. The recession is slowing down because of the gravitational attraction of different parts of the universe to each other, at least in the simpler models.

The expansion of the universe may continue forever, as in the 'open' models, or the expansion may halt at some future time and contraction set in, as in the 'closed' models, in which case the universe will collapse at a finite time later into a space-time singularity with infinite or near infinite density. These possibilities are illustrated in Fig. 1.2. In the Friedmann models the open universes have infinite spatial extent whereas the closed models are finite. This is not necessarily the case for the Lemaitre models. Both the Friedmann and Lemaitre models will be discussed in detail in later chapters.

There is an important piece of evidence apart from the recession of the

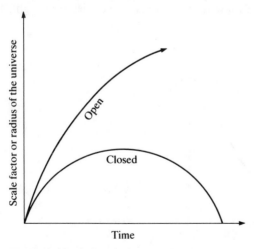

Fig. 1.2. Evolution of the scale factor or radius with time in the open and closed models of the universe.

galaxies that the contents of the universe in the past must have been in a highly compressed form. This is the 'cosmic background radiation', which was discovered by Penzias and Wilson in 1965 and confirmed by many observations later. The existence of this radiation can be explained as follows. As we trace the history of the universe backwards to higher densities, at some stage galaxies could not have had a separate existence, but must have been merged together to form one great continuous mass. Due to the compression the temperature of the matter must have been very high. There is reason to believe, as we shall see, that there must also have been present a great deal of electromagnetic radiation, which at some stage was in equilibrium with the matter. The spectrum of the radiation would thus correspond to a black body of high temperature. There should be a remnant of this radiation, still with black-body spectrum, but corresponding to a much lower temperature. The cosmic background radiation discovered by Penzias, Wilson and others indeed does have a black-body spectrum (Fig. 1.3) with a temperature of about 2.7 K.

Hubble's law implies arbitrarily large velocities of the galaxies as the distance increases indefinitely. There is thus an apparent contradiction with special relativity which can be resolved as follows. The red-shift z is defined as $z = (\lambda_r - \lambda_i)/\lambda_i$, where λ_i is the original wavelength of the radiation given off by the galaxy and λ_r is the wavelength of this radiation when received by us. As the velocity of the galaxy approaches that of light, z tends towards infinity (Fig. 1.4), so it is not possible to *observe* higher velocities than that of light. The distance at which the red-shift of a galaxy becomes infinite is called the *horizon*. Galaxies beyond the horizon are indicated by Hubble's law to have higher velocities than light, but this

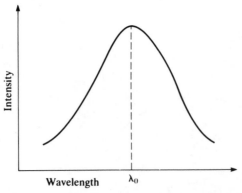

Fig. 1.3. Graph of intensity versus wavelength for black-body radiation. For the cosmic background radiation λ_0 is just under 0.1 cm.

does not violate special relativity because the presence of gravitation radically alters the nature of space and time according to general relativity. It is not as if a material particle is going past an observer at a velocity greater than that of light, but it is space which is in some sense expanding faster than the speed of light. This will become clear when we derive the expressions for the velocity, red-shift, etc. analytically later.

As mentioned earlier, in the open model the universe will expand forever whereas in the closed model there will be contraction and collapse in the future. It is not known at present whether the universe is open or closed. There are several interconnecting ways by which this could be determined. One way is to measure the present average density of the universe and compare it with a certain critical density. If the density is above the critical density, the attractive force of different parts of the universe towards each other will be enough to halt the recession eventually and to pull the galaxies together. If the density is below the critical density, the attractive force is insufficient and the expansion will continue forever. The critical density at any time (this will be derived in detail later) is given by

$$\varepsilon_c = 3H^2/8\pi G, \quad H = \dot{R}/R. \tag{1.3}$$

Here G is Newton's gravitational constant and R is the scale factor which is a function of time; it corresponds to $R'(t)$ of (1.1) and represents

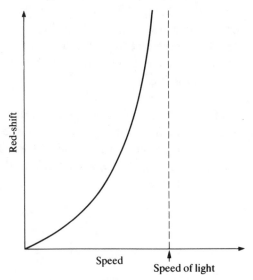

Fig. 1.4. This graph shows the relation between the red-shift (z) and the speed of recession. As z tends to infinity, the speed of recession tends to the speed of light.

the 'size' of the universe in a sense which will become clear later. If t_0 denotes the present time, then the present value of H, denoted by H_0, is called Hubble's constant. That is, $H_0 = H(t_0)$. For galaxies which are not too near nor too far, the velocity v is related to the distance d by Hubble's constant:

$$v = H_0 d. \tag{1.4}$$

(Compare (1.2), (1.3) and (1.4).) The present value of the critical density is thus $3H_0^2/8\pi G$, and is dependent on the value of Hubble's constant. There are some uncertainties in the value of the latter, the likely value being between 50 km s^{-1} and 100 km s^{-1} per million parsecs. That is, a galaxy which is 100 million parsecs distant has a velocity away from us of 5000–10 000 km s^{-1}. For a value of Hubble's constant given by 50 km s^{-1} per million parsecs, the critical density equals about 5×10^{-30} g cm^{-3}, or about three hydrogen atoms per thousand litres of space.

There are several other related ways of determining if the universe will expand forever. One of these is to measure the rate at which the expansion of the universe is slowing down. This is measured by the deceleration parameter, about which there are also uncertainties. Theoretically in the simpler models, in suitable units, the deceleration parameter is half the ratio of the actual density to the critical density. This ratio is usually denoted by Ω. Thus if $\Omega < 1$, the density is subcritical and the universe will expand forever, the opposite being the case if $\Omega > 1$. The present observed value of Ω is somewhere between 0.1 and 2 (the lower limit could be less). In the simpler models the deceleration parameter, usually denoted by q_0, is thus $\frac{1}{2}\Omega$, so that the universe expands forever in these models if $q_0 < \frac{1}{2}$, the opposite being the case if $q_0 > \frac{1}{2}$.

Another way to find out if the universe will expand forever is to determine the precise age of the universe and compare it with the 'Hubble time'. This is the time elapsed since the big bang until now if the rate of expansion had been the same as at present. In Fig. 1.5 if ON denotes the present time (t_0), then clearly PN is $R(t_0)$. If the tangent at P to the curve $R(t)$ meets the t-axis at T at an angle α, then

$$\tan \alpha = PN/NT = \dot{R}(t_0), \tag{1.5}$$

so that

$$NT = PN/\dot{R}(t_0) = R(t_0)/\dot{R}(t_0)$$
$$= H_0^{-1}. \tag{1.6}$$

Thus NT, which is, in fact, Hubble's time, is the reciprocal of Hubble's constant in the units considered here. For the value of 50 km s^{-1} per million parsecs of Hubble's constant, the Hubble time is about 20 billion years. Again in the simpler models, if the universe is older than two-thirds of the Hubble time it will expand forever, the opposite being the case if its age is less than two-thirds of the Hubble time.

Whether the universe will expand forever is one of the most important unresolved problems in cosmology, both theoretically and observationally, but all the above methods of ascertaining this contain many uncertainties.

In this book we shall use the term 'open' to mean a model which expands forever, and 'closed' for the opposite. Sometimes the expression 'closed' is used to mean a universe with a finite volume, but as mentioned earlier, it is only in the Friedman models that a universe has infinite volume if it expands forever, etc.

The standard big-bang model of the universe has had three major successes. Firstly, it predicts that something like Hubble's law of expansion must hold for the universe. Secondly, it predicts the existence of the microwave background radiation. Thirdly, it predicts successfully the formation of light atomic nuclei from protons and neutrons a few minutes after the big bang. This prediction gives the correct abundance ratio for He^3, D, He^4 and Li^7. (We shall discuss this in detail later.) Heavier elements are thought to have been formed much later in the interior of stars.

Certain problems and puzzles remain in the standard model. One of these is that the universe displays a remarkable degree of large-scale homogeneity. This is most evident in the microwave background radiation which is known to be uniform in temperature to about one part in 10 000. The reason this is a puzzle is that soon after the big bang, regions which

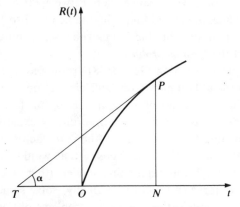

Fig. 1.5. Diagram to define Hubble time.

were well separated could not have communicated with each other or known of each other's existence. Roughly speaking, at a time t after the big bang, light could have travelled only a distance ct since the big bang, so regions separated by a distance greater than ct at time t could not have influenced each other. The fact that microwave background radiation received from all directions is uniform implies that there is uniformity in regions whose separation must have been many times the distance ct (the *horizon distance*) a second or so after the big bang. How did these different regions manage to have the same density, etc.? Of course there is no problem if one simply *assumes* that the uniformity persists up to time $t = 0$, but this requires a very special set of initial conditions. This is known as the *horizon problem*.

Another problem is concerned with the fact that a certain amount of inhomogeneity must have existed in the primordial matter to account for the clumping of matter into galaxies and clusters of galaxies, etc., that we observe today. Any small inhomogeneity in the primordial matter rapidly grows into a large one with gravitational self-interaction. Thus one has to assume a considerable smoothness in the primordial matter to account for the inhomogeneity in the scale of galaxies at the present time. The problem becomes acute if one extrapolates to 10^{-45} s after the big bang, when one has to assume an unusual situation of almost perfect smoothness but not quite absolute smoothness in the initial state of matter. This is known as the *smoothness problem*.

A third problem of the standard big-bang model has to do with the present observed density of matter, which we have denoted by the parameter Ω. If Ω were initially equal to unity (this corresponds to a flat universe) it would stay equal to unity forever. On the other hand, if Ω were initially different from unity, its departure from unity would increase with time. The present value of Ω lies somewhere between 0.1 and 2. For this to be the case the value of Ω would have had to be equal to 1 to one part in 10^{15} a second or so after the big bang, which seems an unlikely situation. This is called the *flatness problem*.

To deal with these problems Alan Guth (1981) proposed a model of the universe, known as the inflationary model, which does not differ from the standard model after a fraction of a second or so, but from about 10^{-45} to 10^{-30} seconds it has a period of extraordinary expansion, or inflation, during which time typical distances (the scale factor) increase by a factor of about 10^{50} more than the increase that would obtain in the standard model. Although the inflationary models (there have been variations of the one put forward by Guth originally) solve some of the problems of the standard models, they throw up problems of their own, which have not all been dealt with in a satisfactory manner. These models will be considered in detail in this book.

The consideration of the universe in the first second or so calls for a great deal of information from the theory of elementary particles, particularly in the inflationary models. This period is referred to as 'the very early universe' and it also provides a testing ground for various theories of elementary particles. These questions will be considered in some detail in a later chapter.

As one extrapolates in time to the very early universe and towards the big bang at $t = 0$, densities become higher and higher and the curvature of space-time becomes correspondingly higher, and at some stage general relativity becomes untenable and one has to resort to the quantum theory of gravitation. However, a satisfactory quantum theory for gravity does not yet exist. Some progress has been made in what is called 'quantum cosmology', in which quantum considerations throw some light on problems to do with initial conditions of the universe. We shall attempt to provide an introduction to this subject in this book.

If the universe is open, that is, if it expands for ever, one has essentially infinite time in the future for the universe to evolve. What will be the nature of this evolution and what will be the final state of the universe? These questions and related ones will be considered in Chapter 10.

2

The Robertson–Walker metric and the Einstein equations

2.1 Summary of general relativity

The Robertson–Walker metric or line-element is fundamental in the standard models of cosmology. The mathematical framework in which the Robertson–Walker metric occurs is that of general relativity. The reader is assumed to be familiar with general relativity but we shall give a brief review here as a reminder of the main results and for the sake of completeness. We shall then go on to derive the Robertson–Walker metric and obtain the Einstein equations for this metric.

General relativity is formulated in a four-dimensional Riemannian space in which points are labelled by a general coordinate system (x^0, x^1, x^2, x^3), often written as x^μ ($\mu = 0, 1, 2, 3$). (Greek indices take values of $0, 1, 2, 3$ and repeated Greek indices are to be summed over these values.) Several coordinate patches may be necessary to cover the whole of space-time. The space has three spatial and one time-like dimension.

Under a coordinate transformation from x^μ to x'^μ (in which x'^μ is, in general, a function of x^0, x^1, x^2, x^3) a contravariant vector field A^μ and a covariant vector field B_μ transform as follows:

$$A'^\mu = \frac{\partial x'^\mu}{\partial x^\nu} A^\nu, \quad B'_\mu = \frac{\partial x^\nu}{\partial x'^\mu} B_\nu, \tag{2.1}$$

and a mixed tensor such as $A^\mu_{\nu\lambda}$ transforms as follows:

$$A'^\mu_{\nu\lambda} = \frac{\partial x'^\mu}{\partial x^\rho} \frac{\partial x^\sigma}{\partial x'^\nu} \frac{\partial x^\tau}{\partial x'^\lambda} A^\rho_{\sigma\tau}, \tag{2.2}$$

etc. All the information about the gravitational field is contained in the second rank covariant tensor $g_{\mu\nu}$ (the number of indices gives the rank of the tensor) called the metric tensor, or simply the metric, which determines the square of the space-time interval ds^2 between infinitesimally separated events or points x^μ and $x^\mu + dx^\mu$ as follows ($g_{\mu\nu} = g_{\nu\mu}$):

$$ds^2 = g_{\mu\nu} \, dx^\mu \, dx^\nu. \tag{2.3}$$

The contravariant tensor corresponding to $g_{\mu\nu}$ is denoted by $g^{\mu\nu}$ and is defined by

$$g_{\mu\nu}g^{\nu\lambda} = \delta_\mu^\lambda. \tag{2.4}$$

where δ_μ^λ is the Kronecker delta, which equals unity if $\lambda = \mu$ (no summation) and zero otherwise. Indices can be raised or lowered by using the metric tensor as follows:

$$A^\mu = g^{\mu\nu}A_\nu, \quad A_\mu = g_{\mu\nu}A^\nu. \tag{2.5}$$

The generalization of ordinary (partial) differentiation to Riemannian space is given by covariant differentiation denoted by a semi-colon and defined for a contravariant and a covariant vector as follows:

$$A^\mu{}_{;\nu} = \frac{\partial A^\mu}{\partial x^\nu} + \Gamma^\mu_{\nu\lambda} A^\lambda, \tag{2.6a}$$

$$A_{\mu;\nu} = \frac{\partial A_\mu}{\partial x^\nu} - \Gamma^\lambda_{\mu\nu} A_\lambda. \tag{2.6b}$$

Here the $\Gamma^\mu_{\nu\lambda}$ are called Christoffel symbols; they have the property $\Gamma^\mu_{\nu\lambda} = \Gamma^\mu_{\lambda\nu}$ and are given in terms of the metric tensor as follows:

$$\Gamma^\mu_{\nu\lambda} = \tfrac{1}{2}g^{\mu\sigma}(g_{\sigma\nu,\lambda} + g_{\sigma\lambda,\nu} - g_{\nu\lambda,\sigma}), \tag{2.7}$$

where a comma denotes partial differentiation with respect to the corresponding variable: $g_{\sigma\nu,\lambda} \equiv \partial g_{\sigma\nu}/\partial x^\lambda$. For covariant differentiation of tensors of higher rank, there is a term corresponding to each contravariant index analogous to the second term in (2.6a) and a term corresponding to each covariant index analogous to the second term in (2.6b) (with a negative sign). Equation (2.7) has the consequence that the covariant derivative of the metric tensor vanishes:

$$g_{\mu\nu;\lambda} = 0, \quad g^{\mu\nu}{}_{;\lambda} = 0. \tag{2.8}$$

This has, in turn, the consequence that indices can be raised and lowered inside the sign for covariant differentiation, as follows:

$$g_{\sigma\mu}A^\mu{}_{;\nu} = A_{\sigma;\nu}, \quad g^{\sigma\mu}A_{\mu;\nu} = A^\sigma{}_{;\nu}. \tag{2.9}$$

Under a coordinate transformation from x^μ to x'^μ the $\Gamma^\mu_{\nu\lambda}$ transform as follows:

$$\Gamma'^{\mu}_{\nu\lambda} = \frac{\partial x'^{\mu}}{\partial x^{\rho}} \frac{\partial x^{\sigma}}{\partial x'^{\nu}} \frac{\partial x^{\tau}}{\partial x'^{\lambda}} \Gamma^{\rho}_{\sigma\tau} + \frac{\partial^2 x^{\sigma}}{\partial x'^{\nu} \partial x'^{\lambda}} \frac{\partial x'^{\mu}}{\partial x^{\sigma}}, \tag{2.10}$$

so that that $\Gamma^{\mu}_{\nu\lambda}$ do not form components of a tensor since the transformation law (2.10) is different from that of a tensor (see (2.2)). At any specific point a coordinate system can always be chosen so that the $\Gamma^{\mu}_{\nu\lambda}$ vanish at the point. From (2.7) it follows that the first derivatives of the metric tensor also vanish at this point. This is one form of the equivalence principle, according to which the gravitational field can be 'transformed away' at any point by choosing a suitable frame of reference. At this point one can carry out a further linear transformation of the coordinates to reduce the metric to that of flat (Minkowski) space:

$$ds^2 = (dx^0)^2 - (dx^1)^2 - (dx^2)^2 - (dx^3)^2, \tag{2.11}$$

where $x^0 = ct$, t being the time and (x^1, x^2, x^3) being Cartesian coordinates. For any covariant vector A_{μ} it can be shown that

$$A_{\mu;\nu;\lambda} - A_{\mu;\lambda;\nu} = A_{\sigma} R^{\sigma}_{\mu\nu\lambda}, \tag{2.12}$$

where $R^{\sigma}_{\mu\nu\lambda}$ is the Riemann tensor defined by

$$R^{\sigma}_{\mu\nu\lambda} = \Gamma^{\sigma}_{\mu\lambda, \nu} - \Gamma^{\sigma}_{\mu\nu, \lambda} + \Gamma^{\alpha}_{\alpha\nu}\Gamma^{\alpha}_{\mu\lambda} - \Gamma^{\sigma}_{\alpha\lambda}\Gamma^{\alpha}_{\mu\nu}. \tag{2.13}$$

The Riemann tensor has the following symmetry properties:

$$R_{\sigma\mu\nu\lambda} = -R_{\mu\sigma\nu\lambda} = -R_{\sigma\mu\lambda\nu}, \tag{2.14a}$$

$$R_{\sigma\mu\nu\lambda} = R_{\nu\lambda\sigma\mu}, \tag{2.14b}$$

$$R_{\sigma\mu\nu\lambda} + R_{\sigma\lambda\mu\nu} + R_{\sigma\nu\lambda\mu} = 0, \tag{2.14c}$$

and satisfies the Bianchi identity:

$$R^{\sigma}_{\mu\nu\lambda;\rho} + R^{\sigma}_{\mu\rho\nu;\lambda} + R^{\sigma}_{\mu\lambda\rho;\nu} = 0. \tag{2.15}$$

The Ricci tensor $R_{\mu\nu}$ is defined by

$$R_{\mu\nu} = g^{\lambda\sigma} R_{\lambda\mu\sigma\nu} = R^{\sigma}_{\mu\sigma\nu}. \tag{2.16}$$

From (2.13) and (2.16) it follows that $R_{\mu\nu}$ is given as follows:

$$R_{\mu\nu} = \Gamma^{\lambda}_{\mu\nu, \lambda} - \Gamma^{\lambda}_{\mu\lambda, \nu} + \Gamma^{\lambda}_{\mu\nu}\Gamma^{\sigma}_{\lambda\sigma} - \Gamma^{\sigma}_{\mu\lambda}\Gamma^{\lambda}_{\nu\sigma}. \tag{2.17}$$

Let the determinant of $g_{\mu\nu}$ considered as a matrix be denoted by g. Then another expression for $R_{\mu\nu}$ is given by the following:

$$R_{\mu\nu} = \frac{1}{(-g)^{1/2}} [\Gamma^\lambda_{\mu\nu}(-g)^{1/2}]_{,\lambda} - [\log(-g)^{1/2}]_{,\mu\nu} - \Gamma^\sigma_{\mu\lambda}\Gamma^\lambda_{\nu\sigma}. \quad (2.18)$$

This follows from the fact that from (2.7) and the properties of matrices one can show that

$$\Gamma^\lambda_{\mu\lambda} = [\log(-g)^{1/2}]_{,\mu}. \quad (2.19)$$

From (2.18) it follows that $R_{\mu\nu} = R_{\nu\mu}$. There is no agreed convention for the signs of the Riemann and Ricci tensors – some authors define these with opposite signs to (2.13) and (2.17). The Ricci scalar R is defined by

$$R = g^{\mu\nu}R_{\mu\nu}. \quad (2.20)$$

By contracting the Bianchi identity on the pair of indices $\mu\nu$ and $\sigma\rho$ (that is, multiplying it by $g^{\mu\nu}$ and $g^{\sigma\rho}$) one can deduce the identity

$$(R^{\mu\nu} - \tfrac{1}{2}g^{\mu\nu}R)_{;\nu} = 0. \quad (2.21)$$

The tensor $G^{\mu\nu} = R^{\mu\nu} - \tfrac{1}{2}g^{\mu\nu}R$ is sometimes called the Einstein tensor.

We are now in a position to write down the fundamental equations of general relativity. These are Einstein's equations given by:

$$R_{\mu\nu} - \tfrac{1}{2}g_{\mu\nu}R = (8\pi G/c^4)T_{\mu\nu}, \quad (2.22)$$

where $T_{\mu\nu}$ is the energy–momentum tensor of the source producing the gravitational field and G is Newton's gravitational constant. For a perfect fluid, $T_{\mu\nu}$ takes the following form:

$$T^{\mu\nu} = (\varepsilon + p)u^\mu u^\nu - pg^{\mu\nu}, \quad (2.23)$$

where ε is the mass-energy density, p is the pressure and u^μ is the four-velocity of matter given by

$$u^\mu = \frac{dx^\mu}{ds}, \quad (2.24)$$

where $x^\mu(s)$ describes the world-line of matter in terms of the proper time $\tau = c^{-1}s$ along the world-line. We shall not be concerned with other forms of the energy–momentum tensor than (2.23). From (2.21) we see that Einstein's equations (2.22) are compatible with the following equation

$$T^{\mu\nu}_{\;\;;\nu} = 0, \quad (2.25)$$

which is the equation for the conservation of mass-energy and momentum.

The equations of motion of a particle in a gravitational field are given by the geodesic equations as follows:

$$\frac{d^2x^\mu}{ds^2} + \Gamma^\mu_{\lambda\nu} \frac{dx^\lambda}{ds} \frac{dx^\nu}{ds} = 0. \tag{2.26}$$

Geodesics can also be introduced through the concept of parallel transfer. Consider a curve $x^\mu(\lambda)$, where x^μ are suitably differentiable functions of the real parameter λ, varying over some interval of the real line. It is readily verified that $dx^\mu/d\lambda$ transforms as a contravariant vector. This is the tangent vector to the curve $x^\mu(\lambda)$. For an arbitrary vector field Y^μ its covariant derivative along the curve (defined along the curve) is $Y^\mu_{;\nu}(dx^\nu/d\lambda)$. The vector field Y^μ is said to be parallelly transported along the curve if

$$Y^\mu_{;\nu} \frac{dx^\nu}{d\lambda} = Y^\mu_{,\nu} \frac{dx^\nu}{d\lambda} + \Gamma^\mu_{\nu\sigma} Y^\sigma \frac{dx^\nu}{d\lambda}$$

$$= \frac{dY^\mu}{d\lambda} + \Gamma^\mu_{\nu\sigma} Y^\sigma \frac{dx^\nu}{d\lambda} = 0. \tag{2.27}$$

The curve is said to be a geodesic curve if the tangent vector is transported parallelly along the curve, that is, putting ($Y^\mu = dx^\mu/d\lambda$ in (2.27)) if

$$\frac{d^2x^\mu}{d\lambda^2} + \Gamma^\mu_{\nu\sigma} \frac{dx^\nu}{d\lambda} \frac{dx^\sigma}{d\lambda} = 0. \tag{2.28}$$

If the tangent vector $dx^\mu/d\lambda$ is time-like everywhere, the curve $x^\mu(\lambda)$ can be taken to be the world-line of a particle and λ the proper time $c^{-1}s$ along the world-line, and in this case (2.28) reduces to (2.26). The former equation has more general applicability, for example, when the curve $x^\mu(\lambda)$ is light-like or space-like, in which case λ cannot be taken as the proper time.

We will now briefly discuss Killing vectors, as these have some relevance for the derivation of the Robertson–Walker metric. In the following we will sometimes write x, y, x' for x^μ, y^μ, x'^μ respectively. A metric $g_{\mu\nu}(x)$ is form-invariant under a transformation from x^μ to x'^μ if $g'_{\mu\nu}(x')$ is the same function of x'^μ as $g_{\mu\nu}(x)$ is of x^μ. For example, the Minkowski metric is form-invariant under a Lorentz transformation. Thus

$$g'_{\mu\nu}(y) = g_{\mu\nu}(y), \quad \text{all } y. \tag{2.29}$$

Therefore

$$g_{\mu\nu}(x) = \frac{\partial x'^\rho}{\partial x^\mu} \frac{\partial x'^\sigma}{\partial x^\nu} g'_{\rho\sigma}(x') = \frac{\partial x'^\rho}{\partial x^\mu} \frac{\partial x'^\sigma}{\partial x^\nu} g_{\rho\sigma}(x'). \tag{2.30}$$

The transformation from x^μ to x'^μ in this case is called an isometry of $g_{\mu\nu}$. Consider an infinitesimal isometry transformation from x^μ to x'^μ defined by

$$x'^\mu = x^\mu + \alpha \xi^\mu(x), \tag{2.31}$$

with α constant and $|\alpha| \ll 1$. Substituting in (2.30) and neglecting terms involving α^2 we arrive at the following equation (see, for example, Weinberg (1972)):

$$g_{\mu\sigma}\frac{\partial \xi^\mu}{\partial x^\rho} + g_{\rho\mu}\frac{\partial \xi^\mu}{\partial x^\sigma} + \frac{\partial g_{\rho\sigma}}{\partial x^\mu}\xi^\mu = 0. \tag{2.32}$$

With the use of (2.6b) and (2.7), (2.32) can be written as follows:

$$\xi_{\sigma;\rho} + \xi_{\rho;\sigma} = 0. \tag{2.33}$$

Equation (2.33) is Killing's equation and a vector field ξ^μ satisfying it is called a Killing vector of the metric $g_{\mu\nu}$. Thus if there exists a solution of (2.33) for a given $g_{\mu\nu}$, then the corresponding ξ^μ represents an infinitesimal isometry of the metric $g_{\mu\nu}$ and implies that the metric has a certain symmetry.

Since (2.33) is covariantly expressed, that is, it is a tensor equation, if the metric has an isometry in one coordinate system, in any transformed coordinate system the transformed metric will also have a corresponding isometry. One can use Killing vectors to derive the Robertson–Walker metric rigorously. However, we will consider this derivation only briefly. Our main derivation will be less rigorous and intuitive, but simple. We now proceed to consider the latter derivation.

2.2   Derivation of the Robertson–Walker metric

As we saw in the last chapter, the universe appears to be homogeneous and isotropic around us on scales of more than a 100 million light years or so, so that on this scale the density of galaxies is approximately the same and all directions from us appear to be equivalent. From these observations one is led to the Cosmological Principle which states that the universe looks the same from all positions in space at a particular time, and that all directions in space at any point are equivalent. This is an intuitive statement of the Cosmological Principle which needs to be made more precise. For example, what does one mean by 'a particular time'? In Newtonian physics this concept is unambiguous. In special relativity the concept becomes well-defined if one chooses a particular inertial frame. In general relativity, however, there are no global inertial frames. To define 'a moment of time' in general relativity which is valid globally, a particular set of circumstances are necessary, which, in fact, are satisfied by a homogeneous and isotropic universe.

To define 'a particular time' in general relativity which is valid globally in this case, we proceed as follows. Introduce a series of non-intersecting space-like hypersurfaces, that is, surfaces any two points of which can be connected to each other by a curve lying entirely in the hypersurface which is space-like everywhere. We make the assumption that all galaxies lie on such a hypersurface in such a manner that the surface of simultaneity of the local Lorentz frame of any galaxy coincides locally with the hypersurface (see Fig. 2.1). In other words, all the local Lorentz frames of the galaxies 'mesh' together to form the hypersurface. Thus the four-velocity of a galaxy is orthogonal to the hypersurface. This series of hypersurfaces can be labelled by a parameter which may be taken as the proper time of any galaxy, that is, time as measured by a clock stationary in the galaxy. As we shall see, this defines a universal time so that a particular time means a given space-like hypersurface on this series of hypersurfaces.

An equivalent description, known as Weyl's postulate (Weyl, 1923) is to assume that the world-lines of galaxies are a bundle or congruence of geodesics in space-time diverging from a point in the (finite or infinitely distant) past, or converging to such a point in the future. These geodesics are non-intersecting, except possibly at a singular point in the past or future or both. There is one and only one such geodesic passing through each regular (that is, a point which is not a singularity) space-time point. This assumption is satisfied to a high degree of accuracy in the actual universe. The deviation from the general motion postulated here is observed to be random and small. The concept of a singular point introduced here will be elucidated at the end of this chapter.

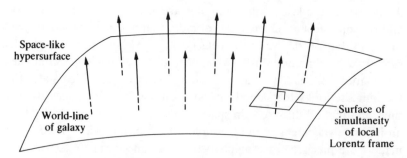

Fig. 2.1. Representation of a typical space-like hypersurface on which galaxies are assumed to lie.

We assume that the bundle of geodesics satisfying Weyl's postulate possesses a set of space-like hypersurfaces orthogonal to them. Choose a parameter t such that each of these hypersurfaces corresponds to $t = $ constant for some constant. The parameter t can be chosen to measure the proper time along a geodesic. Now introduce spatial coordinates (x^1, x^2, x^3) which are constant along any geodesic. Thus for each galaxy the coordinates (x^1, x^2, x^3) are constant. Under these circumstances the metric can be written as follows:

$$ds^2 = c^2 \, dt^2 - h_{ij} \, dx^i \, dx^j, \qquad (i, j = 1, 2, 3), \qquad (2.34)$$

where the h_{ij} are functions of (t, x^1, x^2, x^3) and as usual repeated indices are to be summed over (latin indices take values 1, 2, 3). The fact that the metric given by (2.34) incorporates the properties described above can be seen as follows. Let the world-line of a galaxy be given by $x^\mu(\tau)$, where τ is the proper time along the galaxy. Then according to our assumptions $x^\mu(\tau)$ is given as follows:

$$(x^0 = c\tau, x^1 = \text{constant}, x^2 = \text{constant}, x^3 = \text{constant}). \quad (2.35)$$

From (2.34) and (2.35) we see that the proper time τ along the galaxy is, in fact, equal to the coordinate time t. This is because from (2.35) $dx^i = 0$ along the world-line so that putting $dx^i = 0$ in (2.34) yields $ds = c \, d\tau = c \, dt$, so that $\tau = t$. Clearly a vector along the world-line given by $A^\mu = (c \, dt, 0, 0, 0)$ and the vector $B^\mu = (0, dx^1, dx^2, dx^3)$ lying in the hypersurface $t = $ constant are orthogonal, that is,

$$g_{\mu\nu} A^\mu B^\nu = 0, \qquad (2.36)$$

since $g_{0i} = 0$ $(i = 1, 2, 3)$ in the metric given by (2.34). Further, the world-line given by (2.35) satisfies the geodesic equation

$$\frac{d^2 x^\mu}{ds} + \Gamma^\mu_{\lambda\nu} \frac{dx^\lambda}{ds} \frac{dx^\nu}{ds} = 0. \qquad (2.37)$$

This can be seen from the fact that, from (2.35), we have

$$dx^\mu/ds = (1, 0, 0, 0) \qquad (2.38)$$

so that (2.37) is satisfied if $\Gamma^\mu_{00} = 0$. In fact

$$\Gamma^\mu_{00} = \tfrac{1}{2} g^{\mu\nu}(2g_{\nu 0,0} - g_{00,\nu}). \qquad (2.39)$$

Using the fact that $g^{0i} = 0$ $(i = 1, 2, 3)$ which follows from (2.34), it is readily verified that Γ^μ_{00} given by (2.39) vanishes, so that (2.37) is satisfied and that the world-lines given by (2.35) are indeed geodesics.

The metric given by (2.34) does not incorporate the property that space is homogeneous and isotropic. This form of the metric can be used, with the help of a special coordinate system obtained by singling out a particular typical galaxy, to derive some general properties of the universe without the assumptions of homogeneity and isotropy (see, for example, Raychaudhuri (1955)). We shall not be concerned with this general form (see Chapter 6), but we proceed to consider the form taken by (2.34) when space is homogeneous and isotropic.

The spatial separation on the same hypersurface $t = $ constant of two nearby galaxies at coordinates (x^1, x^2, x^3) and $(x^1 + \Delta x^1, x^2 + \Delta x^2, x^3 + \Delta x^3)$ is

$$d\sigma^2 = h_{ij}\Delta x^i \Delta x^j. \tag{2.40}$$

Consider the triangle formed by three nearby galaxies at some particular time, and the triangle formed by these same galaxies at some later time. By the postulate of homogeneity and isotropy all points and directions on a particular hypersurface are equivalent, so that the second triangle must be similar to the first one and further, the magnification factor must be independent of the position of the triangle in the three-space. It follows that the functions h_{ij} must involve the time coordinate t through a common factor so that ratios of small distances are the same at all times. Thus the metric has the form

$$ds^2 = c^2\,dt^2 - R^2(t)\gamma_{ij}\,dx^i\,dx^j, \tag{2.41}$$

where the γ_{ij} are functions of (x^1, x^2, x^3) only. Consider the three-space given by

$$d\sigma'^2 = \gamma_{ij}\,dx^i\,dx^j. \tag{2.42}$$

We assume this three-space to be homogeneous and isotropic. According to a theorem of differential geometry, this must be a space of constant curvature (see, for example, Eisenhart (1926) or Weinberg (1972)). In such a space the Riemann tensor can be constructed from the metric (and not its derivatives) and constant tensors only. The following three-dimensional fourth rank tensor constructed out of the three-dimensional metric tensor of (2.42) has the correct symmetry properties for the Riemann tensor:

$$^{(3)}R_{ijkl} = k(\gamma_{ik}\gamma_{jl} - \gamma_{il}\gamma_{jk}), \tag{2.43}$$

where k is a constant. One can verify that the three-dimensional Riemann tensor of the space given by (2.42) has the form (2.43) if the γ_{ij} are chosen to be given by the following metric:

$$d\sigma'^2 = (1 + \tfrac{1}{4}kr'^2)^{-2}[(dx^1)^2 + (dx^2)^2 + (dx^3)^2],$$

$$r'^2 = (x^1)^2 + (x^2)^2 + (x^3)^2. \tag{2.44}$$

The metric (2.41) can then be written as follows:

$$ds^2 = c^2\,dt^2 - \frac{R^2(t)(dx^2 + dy^2 + dz^2)}{[1 + \tfrac{1}{4}k(x^2 + y^2 + z^2)]^2}, \tag{2.45}$$

where we have set $x^1 = x$, $x^2 = y$, $x^3 = z$, so that $r'^2 = x^2 + y^2 + z^2$. With $x = r' \sin\theta \cos\phi$, $y = r' \sin\theta \sin\phi$, $z = r' \cos\theta$, (2.45) reduces to the following:

$$ds^2 = c^2\,dt^2 - R^2(t)\left[\frac{dr'^2 + r'^2(d\theta^2 + \sin^2\theta\,d\phi^2)}{(1 + \tfrac{1}{4}kr'^2)^2}\right]. \tag{2.46}$$

The transformation $r = r'/(1 + \tfrac{1}{4}kr'^2)$ yields the standard form of the Robertson–Walker metric, as follows:

$$ds^2 = c^2\,dt^2 - R^2(t)\left[\frac{dr^2}{1 - kr^2} + r^2(d\theta^2 + \sin^2\theta\,d\phi^2)\right]. \tag{2.47}$$

The constant k in (2.47) can take the values $-1, 0, +1$, giving three different kinds of spatial metrics. We will deal with these in detail later.

We will now give a brief discussion of the manner in which the Robertson–Walker metric is derived more rigorously with the help of Killing vectors. A space is said to be homogeneous if there exists an infinitesimal isometry of the metric which can carry any point into any other point in its neighbourhood. From the discussion of Killing vectors it follows that this implies the existence of Killing vectors of the metric which at any point can take all possible values. These remarks can be illustrated by a simple example. Consider the following metric:

$$ds^2 = A(t)\,dt^2 - B(t)\,dx^2 - C(t)\,dy^2 - D(t)\,dz^2, \tag{2.48}$$

where A, B, C, D are functions of the time coordinate t only, and x, y, z are the spatial coordinates. Consider two arbitrary points P and P' with spatial coordinates (a, b, c) and (a', b', c') respectively. Consider now the transformation given by

$$x' = x + a' - a, \quad y' = y + b' - b, \quad z' = z + c' - c. \tag{2.49}$$

This transformation takes the point P to the point P', because when $(x, y, z) = (a, b, c)$, we get $(x', y', z') = (a', b', c')$. On the other hand the new metric is given by

$$ds^2 = A(t)\, dt^2 - B(t)\, dx'^2 - C(t)\, dy'^2 - D(t)\, dz'^2. \qquad (2.50)$$

which has the same form in the new coordinates as (2.48) has in the old coordinates. Thus (2.49) represents an isometry of the metric, which is not just infinitesimal but a finite or a global isometry. Thus the metric (2.48) represents a homogeneous space. In terms of Killing vectors, it is easily verified that the vectors given by $\xi^\mu = (0, 1, 0, 0)$, $\eta^\mu = (0, 0, 1, 0)$ and $\zeta^\mu = (0, 0, 0, 1)$ are all Killing vectors, as are any linear combinations of these with arbitrary constant coefficients. One can thus get Killing vectors which take arbitrary spatial values, which correspond to isometries of the form (2.49).

One can similarly define isotropy in terms of isometries and Killing vectors. A space is isotropic at a point X if there exists an infinitesimal isometry which leaves the point X unchanged but takes any direction at X to any other direction, that is, takes any infinitesimal vector at X to any other one. In terms of Killing vectors, this implies the existence of Killing vectors which vanish at X but whose derivatives can take all possible values, subject to Killing's equation. The metric (2.49), although homogeneous, is not in general isotropic. A space is isotropic if it is isotropic about every point in it. Proceeding along these lines one can derive the Robertson–Walker metric with the use of Killing vectors. We refer the interested reader to Weinberg (1972, Chapter 13) for a detailed discussion of this point.

2.3   Some geometric properties of the Robertson–Walker metric

Consider the Robertson–Walker metric (2.47) when $k = 1$. This yields the universe with positive spatial curvature whose spatial volume is finite, as we shall see. In this case it is convenient to introduce a new coordinate ψ by the relation $r = \sin \psi$, so that the metric (2.47) becomes

$$ds^2 = c^2\, dt^2 - R^2(t)[d\psi^2 + \sin^2 \psi(d\theta^2 + \sin^2 \theta\, d\phi^2)]. \qquad (2.51)$$

Some insight may be gained by embedding the spatial part of this metric in a four-dimensional Euclidean space. In general a three-dimensional Riemannian space with a positive definite metric cannot be embedded in a four-dimensional Euclidean space, but the spatial part of (2.51) can, in fact, be so embedded. Before proceeding to do this, we consider a simple example of embedding, namely, that of the space given by the two-dimensional metric

$$d\sigma'^2 = a^2(d\theta^2 + \sin^2 \theta\, d\phi^2). \qquad (2.52)$$

This, of course, is just the surface of a two-sphere and is represented by the equation $x^2 + y^2 + z^2 = a^2$ in ordinary three-dimensional Euclidean space. This is a trivial example of the embedding of the two-surface given by (2.52). However, a metric such as (2.52) describes the intrinsic properties of the surface and does not depend on its embedding in a higher-dimensional space, although in this simple case it is natural to think in terms of the surface of an ordinary sphere in three dimensions. Turning to (2.51), we write the spatial part as follows:

$$d\sigma^2 = R^2[d\psi^2 + \sin^2\psi(d\theta^2 + \sin^2\theta\, d\phi^2)], \tag{2.53}$$

where we concentrate on a particular time t and regard R as constant. Consider now a four-dimensional Euclidean space with coordinates (w, x, y, z) which are Cartesian-like in that the distance between points given by (w_1, x_1, y_1, z_1) and (w_2, x_2, y_2, z_2) is Σ_{12}, where

$$\Sigma_{12}^2 = (w_1 - w_2)^2 + (x_1 - x_2)^2 + (y_1 - y_2)^2 + (z_1 - z_2)^2. \tag{2.54}$$

Thus the metric in this space is given by

$$d\Sigma^2 = dw^2 + dx^2 + dy^2 + dz^2. \tag{2.55}$$

Consider now a surface in this space given parametrically by

$$w = R\cos\psi, \quad x = R\sin\psi\,\sin\theta\cos\phi,$$
$$y = R\sin\psi\,\sin\theta\sin\phi, \quad z = R\sin\psi\,\cos\theta, \tag{2.56}$$

from which we get

$$w^2 + x^2 + y^2 + z^2 = R^2. \tag{2.57}$$

Evaluating dw, dx, dy, dz in terms of $d\psi$, $d\theta$, $d\phi$ from (2.56) and substituting in (2.55) we get precisely the metric given by (2.53). Just as all points and all directions starting from a point on a two-sphere in a three-dimensional Euclidean space are equivalent, so all points and directions on a three-sphere in a four-dimensional Euclidean space are equivalent. This can be seen from the fact that rotations in the four-dimensional embedding space (which can be affected by a 4×4 orthogonal matrix) can move any point and any direction on the three-sphere into any other point and direction respectively, while leaving unchanged the metric (2.55) and the equation of the three-sphere (2.57). This shows that the metric (2.53), that is, the space $t = $ constant in (2.51), is indeed homogeneous and isotropic.

Consider again a particular time t so that R can be taken as constant in (2.56) and (2.57). Consider the two-surface given by $\psi = $ constant $= \psi_0$, which is a two-sphere, as can be seen from (2.56) and (2.57), whence we get $w = R \cos \psi_0$, and

$$x^2 + y^2 + z^2 = R^2 \sin^2 \psi_0. \tag{2.58}$$

The surface area of this two-sphere is $4\pi R^2 \sin^2 \psi_0$. As ψ_0 ranges from 0 to π, one moves outwards from the 'north pole' (given by $\psi_0 = 0$) of the hypersurface through successive two-spheres of area $4\pi R^2 \sin^2 \psi_0$. The area increases until $\psi_0 = \pi/2$, after which it decreases until it is zero at $\psi_0 = \pi$. The distance from the 'north' to the 'south pole' is $R\pi$. This behaviour is similar to what happens on a two-sphere in a three-dimensional Euclidean space, as illustrated in Fig. 2.2. Suppose the radius of the two-sphere is R' and ψ' denotes the co-latitude. The circumference of the circle on the sphere given by $\psi' = \text{constant} = \psi'_0$ is $2\pi R' \sin \psi'_0$, while the distance of this circle from the north pole O is $R'\psi'_0$. The circumference of this circle reaches a maximum at $\psi'_0 = \pi/2$, after which it decreases until it reaches zero at $\psi'_0 = \pi$, when the distance from the north pole along the surface is $R'\pi$, analogously to the previous case.

In the case of the three-space (2.57), the entire surface is swept by the coordinate range $0 \leqslant \psi \leqslant \pi, 0 \leqslant \theta \leqslant \pi, 0 \leqslant \phi \leqslant 2\pi$. The total volume of the three-space (2.53) is

$$\int (-^{(3)}g)^{1/2} \, \mathrm{d}^3x = \int (R \, \mathrm{d}\psi)(R \sin\psi \, \mathrm{d}\theta)(R \sin\psi \sin\theta \, \mathrm{d}\phi) = 2\pi^2 R^3, \tag{2.59}$$

which is finite. Here $^{(3)}g$ is the determinant of the three-dimensional metric.

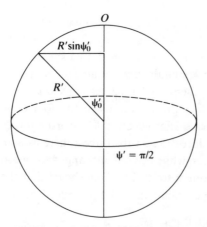

Fig. 2.2. Diagram to illustrate the analogy between the surface of a two-sphere and three-space of positive curvature.

In the case $k = 0$ the spatial metric is given by

$$d\sigma_1^2 = R^2[d\psi^2 + \psi^2(d\theta^2 + \sin^2\theta\, d\phi^2)], \tag{2.60}$$

which is the ordinary three-dimensional Euclidean space. As usual, the transformation

$$x = R\psi \sin\theta \cos\phi, \quad y = R\psi \sin\theta \sin\phi, \quad z = R\psi \cos\theta, \tag{2.61}$$

gives

$$d\sigma_1^2 = dx^2 + dy^2 + dz^2. \tag{2.62}$$

The range of (ψ, θ, ϕ) is $0 \leqslant \psi < \infty$, $0 \leqslant \theta \leqslant \pi$, $0 \leqslant \phi \leqslant 2\pi$, and the spatial volume is infinite. This is also referred to as the universe with zero spatial curvature, as opposed to the case $k = 1$, which has positive spatial curvature.

The case $k = -1$ corresponds to the universe with negative spatial curvature. The spatial part of this metric cannot be embedded in a four-dimensional Euclidean space, but it can be embedded in a four-dimensional Minkowski space. It is, in fact, the space-like surface given by

$$x^2 + y^2 + z^2 - w^2 = -R^2, \tag{2.63}$$

in the Minkowski space with metric

$$ds^2 = dw^2 - dx^2 - dy^2 - dz^2. \tag{2.64}$$

Putting $k = -1$ and $r = \sinh\psi$ in (2.47), we get for the spatial part of this metric the following form

$$d\sigma_2^2 = R^2[d\psi^2 + \sinh^2\psi(d\theta^2 + \sin^2\theta\, d\phi^2)]. \tag{2.65}$$

To see the embedding given by (2.63), (2.64) we transform to a Minkowski space with coordinates (w, x, y, z) given by

$$w = R\cosh\psi, \quad x = R\sinh\psi \sin\theta \cos\phi,$$

$$y = R\sinh\psi \sin\theta \sin\phi, \quad z = R\sinh\psi \cos\theta, \tag{2.66}$$

which gives (2.63) on substitution for w, x, y, z. Evaluating dw, dx, dy, dz from (2.66) in terms of $d\psi, d\theta, d\phi$ and substituting in (2.64) we get the metric (2.65). In this case the surface $w = $ constant given by $\psi = $ constant $= \psi_0$, corresponds, by substituting into (2.63) $w = R\cosh\psi_0$, to the surface of the two-sphere given by

$$x^2 + y^2 + z^2 = R^2 \sinh^2\psi_0. \tag{2.67}$$

The surface of this sphere has area $4\pi R^2 \sinh^2\psi_0$, which keeps on increasing indefinitely as ψ_0 increases. As is clear from the metric (2.65), the 'radius' of this sphere, that is, the distance from the 'centre' given by $\psi = 0$ to the surface given by $\psi = \psi_0$ along $\theta =$ constant and $\phi =$ constant, is $R\psi_0$. Thus the surface area is larger than that of a sphere of radius $R\psi_0$ in Euclidean space. In this case the range of the coordinates (ψ, θ, ϕ) is: $0 \leqslant \psi \leqslant \infty$, $0 \leqslant \theta \leqslant \pi$, $0 \leqslant \phi < 2\pi$. The spatial volume is infinite.

2.4   Some kinematic properties of the Robertson–Walker metric

We have seen that galaxies have fixed spatial coordinates, that is, they are at rest in the coordinate system defined above. Such a system is called comoving. Thus the cosmological 'fluid' is at rest in the comoving frame we have chosen. We now consider the behaviour of a free particle which is travelling with respect to this comoving frame. It is free in the sense that it is affected only by the 'background' cosmological gravitational field and no other forces. This could be a projectile shot out of a galaxy or a light wave (photon) travelling through intergalactic space. Consider the Robertson–Walker metric in the form

$$ds^2 = c^2\, dt^2 - R^2(t)\left[\frac{dr^2}{1 - kr^2} + r^2(d\theta^2 + \sin^2\theta\, d\phi^2)\right]. \qquad (2.68)$$

We write $(x^0, x^1, x^2, x^3) = (ct, r, \theta, \phi)$, so that

$$g_{00} = 1, \quad g_{11} = -R^2(t)/(1 - kr^2),$$

$$g_{22} = -R^2(t)r^2, \quad g_{33} = -R^2(t)r^2\sin^2\theta, \qquad (2.69)$$

the rest of the metric components being zero. Consider a geodesic passing through a typical point P. Without loss of generality we can take the spatial origin of the coordinate system, that is, $r = 0$, to be at the point P. The path of the particle is given by the geodesic equation

$$\frac{du^\mu}{d\lambda} + \Gamma^\mu_{\alpha\beta}u^\alpha u^\beta = 0, \qquad (2.70)$$

where $u^\mu = dx^\mu/d\lambda$, $x^\mu(\lambda)$, being the coordinates of a space-time point on the world-line of the moving particle as a function of the path parameter λ. If the particle is massive, λ can be taken as the proper time s of the particle and if it is a photon, λ is an affine parameter.

Multiply (2.70) by $g_{\sigma\mu}$ and use (2.4), (2.7) to get

$$g_{\sigma\mu}\,(du^\mu/d\lambda) + \tfrac{1}{2}(g_{\sigma\alpha,\beta} + g_{\sigma\beta,\alpha} - g_{\alpha\beta,\sigma})u^\alpha u^\beta = 0. \qquad (2.71)$$

We also have

$$\frac{du_\sigma}{d\lambda} = \frac{d}{d\lambda}(g_{\sigma\mu}u^\mu) = g_{\sigma\mu}\frac{du^\mu}{d\lambda} + g_{\sigma\mu,\rho}u^\rho u^\mu. \tag{2.72}$$

In (2.71) $g_{\sigma\alpha,\beta}u^\alpha u^\beta = g_{\sigma\beta,\alpha}u^\alpha u^\beta$, so that if we eliminate from this equation the term $g_{\sigma\mu}\,du^\mu/d\lambda$ with the use of (2.72), we arrive at the following equation

$$du_\sigma/d\lambda - \tfrac{1}{2}g_{\alpha\beta,\sigma}u^\alpha u^\beta = 0. \tag{2.73}$$

Equation (2.73) tells us that if the metric components are independent of a particular coordinate x^σ, then the covariant component u_σ is constant along the geodesic. Consider the component $\sigma = 3$, so that we are referring to $x^3 = \phi$. Since the metric components (2.69) are independent of ϕ, we have $du_3/d\lambda = 0$, so that u_3 is constant along the geodesic. But

$$u_3 = g_{33}u^3 = -R^2(t)r^2(\sin^2\theta)u^3, \tag{2.74}$$

so that $u_3 = 0$ at the point P where $r = 0$. Thus $u_3 = 0$ along the geodesic and so $u^3 = d\phi/d\lambda = 0$ as well, so ϕ is constant along the geodesic.

Consider (2.73) for $\sigma = 2$:

$$du_2/d\lambda - \tfrac{1}{2}g_{\alpha\beta,2}u^\alpha u^\beta = 0. \tag{2.75}$$

The only component of $g_{\alpha\beta}$ which depends on $x^2 = \theta$ is g_{33}, but the contribution of the corresponding term to (2.75) vanishes since $u^3 = 0$. Thus $du_2/d\lambda = 0$, so u_2 is constant along the geodesic. Again

$$u_2 = g_{22}u^2 = -R^2(t)r^2u^2, \tag{2.76}$$

which vanishes at P $(r = 0)$, and so u_2 is zero along the geodesic, as is u^2, so that θ is also constant along the geodesic.

To proceed further we concentrate on the case $k = 0$ in (2.68) and (2.69). We leave it as an exercise for the reader to extend the following analysis to the cases $k = +1, -1$. In these two cases it is helpful to transform the coordinate r to ψ given by $r = \sin\psi$, $r = \sinh\psi$ respectively, as in (2.53), (2.65). We return to (2.73) with $\sigma = 1$:

$$du_1/d\lambda - \tfrac{1}{2}g_{\alpha\beta,1}u^\alpha u^\beta = 0. \tag{2.77}$$

We have $u^2 = u^3 = 0$, while g_{00} and g_{11} are independent of r (recall that $k = 0$). Thus $du_1/d\lambda = 0$ so that u_1 is constant along the geodesic:

$$u_1 = g_{11}u^1 = -R^2(t)\frac{dr}{ds} = \text{constant}, \tag{2.78}$$

where we have taken the parameter λ to be the proper time s. In the metric (2.68) we can set $d\theta = d\phi = 0$ (since θ and ϕ are constant along the geodesic) to get

$$ds^2 = c^2\, dt^2 - R^2(t)\, dr^2 = c^2\, dt^2 - dl^2 = dt^2(c^2 - v^2), \qquad (2.79)$$

where dl is the element of spatial distance and $v = dl/dt$ is the velocity of the particle in the comoving frame, assuming it to be a massive particle of mass m. The momentum of the particle is given as follows:

$$q = m\,(dl/ds)c = mv/(1 - v^2/c^2)^{1/2}. \qquad (2.80)$$

Combining (2.78), (2.79), (2.80) we get

$$qR(t) = \text{constant along the geodesic.} \qquad (2.81)$$

The above analysis can also be applied to the case of a photon, in which case, since the energy q_0 and the momentum q of the photon are related by $q_0 = cq$, we have

$$q_0 R(t) = \text{constant along the geodesic.} \qquad (2.82)$$

Since the energy of the photon is proportional to its frequency v, we get

$$vR(t) = \text{constant along the geodesic.} \qquad (2.83)$$

Consider a photon emitted at time t_1 with frequency v_1 which is observed at the point P at time t_0 with frequency v_0. From (2.83) we get

$$v_0 R(t_0) = v_1 R(t_1). \qquad (2.84)$$

This can be written as

$$1 + z = R(t_0)/R(t_1), \qquad (2.85)$$

where $z = (\lambda_0 - \lambda_1)/\lambda_1$ is the fractional change in the wavelength; λ_0, λ_1 being the wavelengths corresponding to the frequencies v_0, v_1 (with $v_0\lambda_0 = v_1\lambda_1 = c$). The number z is always observed to be positive, at least for distant galaxies, indicating a shift in the visible spectrum towards red, so that z is referred to as the 'red-shift'. We will come back to (2.85) later, but now we discuss another derivation of this relation.

The light ray follows a path given by $ds = 0$, which, with the use of (2.79), yields the following relation

$$\int_{t_1}^{t_0} c\,(dt/R(t)) = \int_0^{r_1} dr = r_1, \qquad (2.86)$$

assuming the emitting galaxy to be at $r = r_1$. If the next wave train leaves the galaxy at $t_1 + \delta t_1$ and arrives at $t_0 + \delta t_0$, (2.86) implies

$$\int_{t_1}^{t_0} c \, dt/R(t) = \int_{t_1 + \delta t_1}^{t_0 + \delta t_0} c \, dt/R(t). \tag{2.87}$$

Assuming δt_0, δt_1 to be small compared to t_0, t_1, (2.87) can be approximated as follows

$$\delta t_1/R(t_1) = \delta t_0/R(t_0). \tag{2.88}$$

Since the frequency is inversely proportional to the time interval in which the wave train is emitted, we get (2.84) again.

Without any further consideration the function $R(t)$ which occurs in the Robertson–Walker metric can be *any* function of the time t. From (2.85) we see, since z is observed not to be zero, that the function $R(t)$ is not just a constant. To determine this function we must resort to dynamics, which are provided by Einstein's equations. Before considering these, in the next chapter, we discuss some further properties of the Robertson–Walker metric which are independent of what form the function $R(t)$ takes. These properties may be referred to as kinematic properties.

As indicated in Chapter 1, the first evidence of a systematic red-shift in the spectra of light coming from distant galaxies was found by Hubble. He analysed the data on frequency shifts obtained earlier by Slipher and others and found a linear relationship between the red-shift z and the distance l. He interpreted the red-shift as being due to the recessional velocity of the galaxies. The approximate argument, which is valid if the values of the red-shifts are not high, goes as follows. Let δt_1 of the earlier discussion following (2.86) represent the time interval during which successive wave crests leave the source at $r = r_1$, and let δt_0 be the interval during which these wave crests are received by the observer. If the source is moving away from the observer with velocity v, during the time the two consecutive wave crests are emitted the source moves a distance $v\delta t_1$. Because of this movement, the time interval in which the crests reach the observer is increased by an amount $v\delta t_1/c$. Thus we have

$$\delta t_0 = \delta t_1 + v\delta t_1/c. \tag{2.89}$$

The wavelengths of the emitted and observed light are given as follows:

$$\lambda_1 = c\delta t_1, \quad \lambda_0 = c\delta t_0. \tag{2.90}$$

From (2.89) and (2.90) it follows that

$$\lambda_0/\lambda_1 = \delta t_0/\delta t_1 = 1 + v/c = 1 + z. \tag{2.91}$$

Thus $z = v/c$. This is true if the velocity is small compared to the speed of light. From (2.85) and (2.91) we get

$$v = cz = c(t_0 - t_1)\dot{R}(t_1)/R(t_1), \tag{2.92}$$

where we have assumed $t_0 - t_1$ to be small and expanded $R(t)$ about $t = t_1$, with $\dot{R}(t) \equiv dR(t)/dt$. Again if $t_0 - t_1$ is small the t_1 in the arguments of R and $\dot{R}$ in (2.92) can be replaced by t_0. With the use of similar approximations, we derive the following relations between the coordinate distance r_1 and the distance l of the galaxy:

$$r_1 = c(t_0 - t_1)/R(t_0), \tag{2.93}$$

$$l = r_1 R(t_0) = c(t_0 - t_1). \tag{2.94}$$

With the use of (2.92), (2.93) and (2.94) we finally get Hubble's law, as follows

$$v = cz = H_0 l, \quad H_0 = \dot{R}(t_0)/R(t_0). \tag{2.95}$$

There are many uncertainties in the exact determination of Hubble's constant, H_0, some of which we shall discuss later in the book. One of the best values available is that of Sandage and Tamman (1975), as follows:

$$H_0 = (50.3 \pm 4.3) \text{ km s}^{-1} \text{ Mpc}^{-1}. \tag{2.96}$$

Here Mpc stands for megaparsec, which is approximately 3.26 million light years.

As mentioned already, the formula (2.95) holds only when the red-shift is small. We should expect departures from this linear Hubble's law if the red-shift is not small. To this end, we expand $R(t)$ in a Taylor series about the *present* epoch t_0, as follows:

$$\begin{aligned} R(t) &= R[t_0 - (t_0 - t)] \\ &= R(t_0) - (t_0 - t)\dot{R}(t_0) + \tfrac{1}{2}(t_0 - t)^2 \ddot{R}(t_0) - \cdots \\ &= R(t_0)[1 - (t_0 - t)H_0 - \tfrac{1}{2}(t_0 - t)^2 q_0 H_0^2 - \cdots], \end{aligned} \tag{2.97}$$

with

$$q_0 = -\ddot{R}(t_0)R(t_0)/\dot{R}^2(t_0). \tag{2.98}$$

With the use of (2.86) with a minor adjustment of sign we get

$$\begin{aligned} r &= \int_t^{t_0} c\,dt/R(t) = \int_t^{t_0} c\,dt/\{R(t_0)[1 - (t_0 - t)H_0 - \cdots]\} \\ &= cR^{-1}(t_0)[(t_0 - t) + \tfrac{1}{2}(t_0 - t)^2 H_0 + \cdots]. \end{aligned} \tag{2.99}$$

Here r is the coordinate radius of the galaxy under consideration. The first term in the last expression in (2.99) gives (2.93). With the use of the first part of (2.94), namely, $l = rR(t_0)$, we can invert (2.99) to obtain $t_0 - t$ in terms of l as follows

$$t_0 - t = l/c - \tfrac{1}{2}H_0l^2/c^2. \qquad (2.100)$$

From (2.85) and (2.97) we can find z up to second order in $t_0 - t$ as follows:

$$z = [1 - (t_0 - t)H_0 - \tfrac{1}{2}(t_0 - t)^2 q_0 H_0^2 - \cdots]^{-1} - 1$$
$$= (t_0 - t)H_0 + (t_0 - t)^2(\tfrac{1}{2}q_0 + 1)H_0^2 + \cdots. \qquad (2.101)$$

We now substitute for $t_0 - t$ from (2.100) into (2.101) to obtain a relation for the red-shift z in terms of the distance l.

$$z = H_0l/c + \tfrac{1}{2}(1 + q_0)H_0^2l^2/c^2 + O(H_0^3l^3). \qquad (2.102)$$

Thus from the observed red-shifts it is possible to determine the parameters H_0 and q_0 if an independent estimate can be obtained for the distance. The parameter q_0 is referred to as the deceleration parameter, as it indicates by how much the expansion of the universe is slowing down.

2.5   The Einstein equations for the Robertson–Walker metric

In this section we derive the Einstein equations given by (2.22) for the Robertson–Walker metric, in which the matter is in the form of a perfect fluid of mass-energy density ε and pressure p, so that the energy–momentum tensor is given by (2.23), with $u^\mu = (1, 0, 0, 0)$, as we are in comoving coordinates.

The metric components and Christoffel symbols which are non-zero are given as follows (recall that $(x^0, x^1, x^2, x^3) = (ct, r, \theta, \phi)$):

$$g_{00} = 1, \quad g_{11} = -R^2/(1 - kr^2), \quad g_{22} = -r^2R^2,$$
$$g_{33} = \sin^2\theta R^2 \qquad (2.103)$$

$$g^{00} = 1, \quad g^{11} = -(1 - kr^2)/R^2, \quad g^{22} = -(rR)^{-2},$$
$$g^{33} = -(r\sin\theta\, R)^{-2} \qquad (2.104)$$

We put the Christoffel symbols $\Gamma^\mu_{\nu\lambda}$ in four groups according to the values

0, 1, 2, 3 of the index μ, as follows:

$$\Gamma^0_{11} = c^{-1}R\dot{R}/(1 - kr^2), \quad \Gamma^0_{22} = c^{-1}r^2 R\dot{R},$$
$$\Gamma^0_{33} = c^{-1}r^2 \sin^2 \theta R\dot{R}, \tag{2.105a}$$

$$\Gamma^1_{01} = c^{-1}\dot{R}/R, \quad \Gamma^1_{11} = kr/(1 - kr^2), \quad \Gamma^1_{22} = -r(1 - kr^2),$$
$$\Gamma^1_{33} = -r(1 - kr^2)\sin^2 \theta, \tag{2.105b}$$

$$\Gamma^2_{02} = c^{-1}\dot{R}/R, \quad \Gamma^2_{12} = 1/r, \quad \Gamma^2_{33} = -\sin \theta \cos \theta, \tag{2.105c}$$

$$\Gamma^3_{03} = c^{-1}\dot{R}/R, \quad \Gamma^3_{13} = 1/r, \quad \Gamma^3_{23} = \cot \theta. \tag{2.105d}$$

We next substitute the Christoffel symbols into (2.17) or (2.18) to get the following non-zero components of the Ricci tensor $R_{\mu\nu}$ (note that r is dimensionless while $R(t)$ has the dimension of length).

$$R_{00} = -3\ddot{R}/R, \tag{2.106a}$$

$$R_{11} = (R\ddot{R} + 2\dot{R}^2 + 2c^2 k)/(1 - kr^2), \tag{2.106b}$$

$$R_{22} = r^2(R\ddot{R} + 2\dot{R}^2 + 2c^2 k), \tag{2.106c}$$

$$R_{33} = r^2 \sin^2 \theta(R\ddot{R} + 2\dot{R}^2 + 2c^2 k). \tag{2.106d}$$

It is unfortunate that the same letter is normally used for the scale factor $R(t)$ as for the Ricci scalar (see (2.20)), but it should be clear from the context which is meant. The Ricci scalar can be evaluated with the use of (2.106a)–(2.106d) as follows:

$$g^{\mu\nu}R_{\mu\nu} = -6(R\ddot{R} + \dot{R}^2 + c^2 k)/R^2. \tag{2.107}$$

We are now in a position to write down the Einstein equations (2.22), noting that the covariant components of the four-velocity are the same as the contravariant ones: $u_\mu = (1, 0, 0, 0)$, so that the non-zero components of $T_{\mu\nu}$ are:

$$T_{00} = \varepsilon, \quad T_{11} = pR^2/(1 - kr^2), \quad T_{22} = pr^2 R^2,$$
$$T_{33} = pr^2(\sin^2 \theta)R^2. \tag{2.108}$$

The 00- and 11-components of (2.22) can be written as follows:

$$3(\dot{R}^2 + c^2 k) = 8\pi G\varepsilon R^2/c^2, \tag{2.109a}$$

$$2R\ddot{R} + \dot{R}^2 + kc^2 = -8\pi GpR^2/c^2. \tag{2.109b}$$

The 00-component of (2.22) has been multiplied by R^2 to get (2.109a), while the 11-component has been multiplied by $kr^2 - 1$ to get (2.109b). The 22- and 33-components of (2.22) yield equations which are equivalent to (2.109b).

A useful consequence of (2.109a) and (2.109b) can be obtained by considering the equation of conservation of mass-energy given by (2.25). A generalization of (2.6a) implies that (2.25) can be written as follows:

$$T^{\mu\nu}{}_{,\nu} + \Gamma^{\mu}_{\nu\sigma}T^{\sigma\nu} + \Gamma^{\nu}_{\nu\sigma}T^{\mu\sigma} = 0. \tag{2.110}$$

With the non-zero contravariant components $T^{\mu\nu}$ given as follows:

$$T^{00} = \varepsilon, \quad T^{11} = p(1 - kr^2)/R^2, \quad T^{22} = p/(rR)^2,$$

$$T^{33} = p/[r(\sin\theta)R]^2, \tag{2.111}$$

and with the use of the Christoffel symbols (2.105a)–(2.105d), (2.110) can be written as follows:

$$\dot{\varepsilon} + 3(p + \varepsilon)\dot{R}/R = 0, \tag{2.112}$$

which comes from the $\mu = 0$ component of (2.110), the other components being satisfied identically. Equation (2.112) is, of course, a consequence of (2.109a) and (2.109b) and can be derived from these by first evaluating $\dot{\varepsilon}$ from (2.109a) and using (2.109b) to eliminate $\ddot{R}$.

3

The Friedmann models

3.1 Introduction

At the end of the last chapter we derived the Einstein equations for the Robertson–Walker metric with the energy–momentum tensor as that of a perfect fluid in which the matter is at rest in the local frame. While the Robertson–Walker metric incorporates the symmetry properties and the kinematics of space-time, the Einstein equations provide the dynamics, that is, the manner in which the matter, and the space-time in turn, are affected by the forces present in the universe.

We rewrite (2.109a) and (2.109b) as follows. First we eliminate $\dot{R}^2$ from (2.109b) to get the following equation:

$$\ddot{R} = -(4\pi G/3)(\varepsilon + 3p)R/c^2. \tag{3.1}$$

Next we write (2.109a) for the three different values of k: $-1, 0, 1$.

$$\dot{R}^2 = c^2 + (8\pi G/3)\varepsilon R^2/c^2, \tag{3.2a}$$

$$\dot{R}^2 = (8\pi G/3)\varepsilon R^2/c^2, \tag{3.2b}$$

$$\dot{R}^2 = -c^2 + (8\pi G/3)\varepsilon R^2/c^2. \tag{3.2c}$$

For any one of the three values of k, we have two equations for the three unknown functions R, ε, p. We need one more equation, which is provided by the equation of state, $p = p(\varepsilon)$, in which the pressure is given as a function of the mass–energy density. With the equation of state given, the problem is determinate and the three functions R, ε, p can be worked out completely. Models of the universe which are determined in this way are referred to as Friedmann models, after the Russian mathematician A. A. Friedmann (1888–1925) who was the first to study these models.

Some information can be obtained about the function $R(t)$ without solving the equations explicitly, if one makes a few reasonable assumptions about the pressure and density. For example, if we assume that

$\varepsilon + 3p$ remains positive, then from (3.1) we see that the 'acceleration' $\ddot{R}/R$ is negative. Let the present time be denoted by $t = t_0$. Now $R(t_0) > 0$ (by definition) and $\dot{R}(t_0)/R(t_0) > 0$ (because we see red-shifts, not blue-shifts – see (2.92)); it follows that the curve $R(t)$ must be concave downwards (towards the t-axis – see Fig. 3.1). It is also clear from the figure that the curve $R(t)$ must reach the t-axis at a time which is closer to the present time than the time at which the tangent to the point $(t_0, R(t_0))$ reaches the t-axis. We refer to the time at which $R(t)$ reaches the t-axis as $t = 0$. Thus at a finite time in the past, namely $t = 0$, we have

$$R(0) = 0. \tag{3.3}$$

The point $t = 0$ can reasonably be called the beginning of the universe. Clearly, the point at which the tangent meets the t-axis is the point at which $R(t)$ would have been zero if the expansion had been uniform, that is, if $\dot{R}$ was constant and $\ddot{R} = 0$. The time elapsed from that point till the present time is $R(t_0)/\dot{R}(t_0) = H_0^{-1}$ (see the discussion on page 8). Thus, since, in fact, $\ddot{R}$ is negative for $0 < t < t_0$, it follows that the age of the universe must be less than the Hubble time:

$$t_0 < H_0^{-1}. \tag{3.4}$$

Adding $\dot{p}$ to both sides of (2.112) and multiplying the resulting equation by R^3, we get the following equation:

$$\dot{p}R^3 = \frac{\mathrm{d}}{\mathrm{d}t}[R^3(\varepsilon + p)]. \tag{3.5}$$

We multiply (3.5) by $\dot{R}^{-1}$ and transform the derivative with respect to t

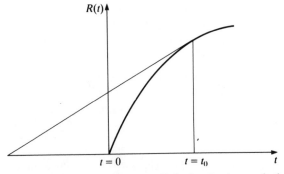

Fig. 3.1. Diagram to illustrate (3.4), that is, the result that the age of the universe is less than the Hubble time.

to a derivative with respect to R, to arrive at the following equation:

$$\frac{d}{dR}(\varepsilon R^3) = -3pR^2.$$ (3.6)

From this equation we see that as long as the pressure p remains positive, the density ε must decrease with increasing R at least as fast as R^{-3}. This is because if the pressure is zero in (3.6), the density varies exactly as R^{-3}, and with a negative right hand side (for positive pressure), the density must decrease faster than R^{-3}. Thus as R tends to infinity, the quantity εR^2 vanishes at least as fast as R^{-1}. We see that in the cases $k = -1$ and $k = 0$, given respectively by (3.2a) and (3.2b), $\dot{R}^2$ remains positive definite so that $R(t)$ keeps on increasing. From (3.2a) we clearly get the result

$$R(t) \to ct \quad \text{as} \quad t \to \infty; \quad k = -1.$$ (3.7)

For $k = 0$ also, $R(t)$ goes on increasing, but more slowly than t. In the case $k = +1$, given by (3.2c), $\dot{R}^2$ becomes zero when εR^2 reaches the value $3c^4/8\pi G$. Since $\ddot{R}$ is negative definite, the curve $R(t)$ must continue to be concave towards the t-axis, so that $R(t)$ begins to decrease, and must reach $R(t) = 0$ at some finite time in the future (the time $t = t_1$ in Fig. 3.2). The three cases $k = -1, 0, +1$ are illustrated in Fig. 3.2.

In (2.96) we have mentioned approximately $50 \text{ km s}^{-1} \text{ Mpc}^{-1}$ as a reasonable value of Hubble's constant. To find the corresponding Hubble time, we merely have to determine the time taken to traverse 1 Mpc at a speed of 50 km s^{-1}. Since there are about 3×10^{19} km in a megaparsec and about 3×10^7 s in a year, we readily see that the value of 50 for H_0

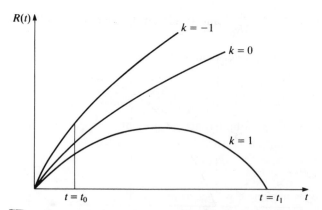

Fig. 3.2. The behaviour of the curve $R(t)$ for the three values $-1, 0, +1$ of k. The time $t = t_0$ is the present time and $t = t_1$ the time at which $R(t)$ reaches zero again for $k = +1$.

in the above units corresponds to a Hubble time of about 20 billion years. Similarly, a value of 100 for H_0 gives a Hubble time of approximately 10 billion years.

Recalling that $H_0 = \dot{R}(t_0)/R(t_0)$ (see (2.95)), the three relations (3.2a)–(3.2c) (that is, (2.109a)) can be written as follows:

$$3c^2 H_0^2/(8\pi G) = -3kc^4/(8\pi G R_0^2) + \varepsilon_0. \tag{3.8}$$

where $R_0 = R(t_0)$. Let us denote the left-hand side of (3.8) by ε_c, and call it the critical density, ε_0 being the present value of the density. From (3.8) we see that if ε_0 is less than ε_c, then k is negative, whereas if ε_0 is greater than ε_c, then k is positive. From the above discussion (see Fig. 3.2) we see that the universe will expand forever if the present density is below the critical density, and it will stop expanding and collapse to zero $R(t)$ at some time in the future if the present density is above the critical density. With $G = 6.67 \times 10^{-8}$ dyne cm^2 g^{-2}, the value of the critical density can be written as follows:

$$\varepsilon_c/c^2 \equiv 3H_0^2/(8\pi G) = 4.9 \times 10^{-30}(H_0/50 \text{ km s}^{-1} \text{ Mpc}^{-1})^2 \text{ g cm}^{-3}. \tag{3.9}$$

Thus if H_0 has the value 50 in the usual units, the critical density is approximately five times 10^{-30} g cm^{-3}, or, since the proton mass is about 1.67×10^{-24} g, about three hydrogen atoms in every thousand litres of space, as mentioned in Chapter 1.

Recalling the definition of the deceleration parameter q_0 (see (2.98)), and denoting by a subscript zero all quantities evaluated at the present epoch $t = t_0$, from (3.1) we get

$$\varepsilon_0 + 3p_0 = -3\ddot{R}_0 c^2/(4\pi G R_0) = (3/4\pi G)q_0 H_0^2 c^2. \tag{3.10}$$

We next eliminate ε_0 between (3.8) and (3.10) to get the following expression for p_0:

$$p_0 = -(8\pi G)^{-1}[kc^2/R_0^2 + H_0^2(1 - 2q_0)]c^2. \tag{3.11}$$

Observationally it is found that the present universe is dominated by non-relativistic matter, that is,

$$p_0 \ll \varepsilon_0, \tag{3.12}$$

so that if p_0 is negligible we get from (3.11)

$$c^2 k/R_0^2 = (2q_0 - 1)H_0^2. \tag{3.13}$$

From (3.8), (3.9) and (3.13) we then get the following simple relation between the ratio of the present density to the critical density and the deceleration parameter:

$$\varepsilon_0/\varepsilon_c = 2q_0. \tag{3.14}$$

We see from (3.14) or directly from (3.13) that the universe is open if q_0 is less than $\frac{1}{2}$ and closed if it is greater than $\frac{1}{2}$. If the present density is exactly equal to the critical density or if q_0 is exactly equal to $\frac{1}{2}$ (together with the assumption of zero pressure in the latter case), we have $k = 0$ and the universe is open (see Fig. 3.2).

3.2   Exact solution for zero pressure

As we have noted, observationally the pressure seems to be negligible compared to the mass-energy density. We shall discuss this further later, but for the present we set $p = 0$, because this yields an exact solution for all time, and, although it may not be accurate, especially for the early epoch of the universe, it provides a useful model. In this case (3.6) can be integrated at once to yield the following equation:

$$\varepsilon/\varepsilon_0 = (R_0/R)^3. \tag{3.15}$$

We eliminate ε_0 and k/R_0^2 with the use of (3.10) (recalling that $p_0 = 0$) and (3.13), and use (3.15) to write (2.109a) as follows:

$$(\dot{R}/R_0)^2 = H_0^2(1 - 2q_0 + 2q_0R_0/R). \tag{3.16}$$

The solution of this equation can be expressed as an integral, giving t in terms of R, as follows:

$$t = H_0^{-1} \int_0^R (1 - 2q_0 + 2q_0R_0/R')^{-1/2} \, dR'/R_0, \tag{3.17}$$

with $t = 0$ being the value of t for which $R(t) = 0$. In particular, the present age of the universe is obtained by taking R_0 as the upper limit in the integral in (3.17). This age can be expressed in terms of H_0 and q_0, both of which are observational parameters, by changing the variable of integration to $w = R'/R_0$, as follows:

$$t_0 = H_0^{-1} \int_0^1 (1 - 2q_0 + 2q_0/w)^{-1/2} \, dw. \tag{3.18}$$

This relation holds for all three values of k, but with the assumption of zero pressure. It is clear that for any positive q_0, the present age t_0 given by (3.18) must satisfy the inequality (3.4). We now consider explicitly three different cases, denoted by (i), (ii) and (iii) below.

(i) $k = +1, \varepsilon_0 > \varepsilon_c$.

From (3.14) we see that this case corresponds to $q_0 > \frac{1}{2}$. In this case the integral in (3.17) can be integrated by the following substitution:

$$1 - \cos \theta = (q_0 R_0)^{-1}(2q_0 - 1)R', \tag{3.19}$$

the resulting equation being given by the following:

$$H_0 t = q_0(2q_0 - 1)^{-3/2}(\theta - \sin \theta). \tag{3.20}$$

After the integration the R' in (3.19) can be replaced by $R(t)$. Equations (3.19) and (3.20) then imply that the curve $R(t)$ is a cycloid. From the left-hand side of (3.19) it is clear that $R(t)$ increases from zero at $\theta = 0$ to its maximum value at $\theta = \pi$, and then decreases steadily until it reaches zero again at $\theta = 2\pi$. The maximum value of $R(t)$ occurs at the time T_m given by

$$T_m = \pi q_0 H_0^{-1}(2q_0 - 1)^{-3/2}, \quad R(T_m) = 2q_0(2q_0 - 1)^{-1}R_0. \tag{3.21}$$

When $R(t)$ returns to zero again, $t = 2T_m$. The present value of θ, θ_0, is given by setting R' equal to R_0 in (3.19), so that

$$\cos \theta_0 = q_0^{-1} - 1. \tag{3.22}$$

Substituting this into (3.20) we get the present age of the universe as

$$t_0 = H_0^{-1}q_0(2q_0 - 1)^{-3/2}[\cos^{-1}(q_0^{-1} - 1) - q_0^{-1}(2q_0 - 1)^{1/2}]. \tag{3.23}$$

If, for example, $q_0 = \frac{2}{3}$, so that $\theta_0 = \pi/3$, and if H_0 is 50 in the units used earlier, so that H_0^{-1} is about 20 billion years, from (3.23) we readily see that t_0 is then approximately 12.3 billion years. In this case T_m is about 218 billion years so that the whole life cycle of the universe is about 436 billion years.

(ii) $k = 0, \varepsilon_0 = \varepsilon_c$.

From (3.14) we see that this case corresponds to $q_0 = \frac{1}{2}$. The integral (3.17) is readily evaluated to yield

$$R(t)/R_0 = (3H_0 t/2)^{2/3}. \tag{3.24}$$

The age of the universe is given by

$$t_0 = \tfrac{2}{3}H_0^{-1}, \tag{3.25}$$

so that for the value 50 for H_0, the age is approximately 13.3 billion years. This case is known as the *Einstein–de Sitter model.*

(iii) $k = -1, \varepsilon_0 < \varepsilon_c.$

From (3.14) it follows that this is the case $q_0 < \tfrac{1}{2}$. The analysis of case (i) above can be taken over if we consider θ to be imaginary and set $\theta = iu$. Equation (3.20) then becomes

$$H_0 t = q_0(1 - 2q_0)^{-3/2}(\sinh u - u), \tag{3.26}$$

where u is given by

$$\cosh u - 1 = (q_0 R_0)^{-1}(1 - 2q_0)R(t). \tag{3.27}$$

As in case (ii), $R(t)$ increases without limit. For large t and u these two variables are related approximately as

$$t = H_0^{-1} q_0(1 - 2q_0)^{-3/2}\exp(u), \tag{3.28}$$

so that as t tends to infinity

$$R(t)/R_0 \to \tfrac{1}{2}q_0(1 - 2q_0)^{-1} \exp(u) \to \tfrac{1}{2}(1 - 2q_0)^{1/2}H_0 t. \tag{3.29}$$

The present value of u, u_0, is obtained by setting R equal to R_0 in (3.27) and is given as follows:

$$\cosh u_0 = q_0^{-1} - 1. \tag{3.30}$$

Substituting this into (3.26), we get the age of the universe as follows:

$$t_0 = H_0^{-1}[(1 - 2q_0)^{-1} - q_0(1 - 2q_0)^{-3/2} \cosh^{-1}(q_0^{-1} - 1)]. \tag{3.31}$$

The mass density of the visible matter, that is, the matter that is contained within the galaxies, is between a tenth and a fifth of the critical density for any reasonable value of Hubble's constant. In this case, if one takes as an example the value 0.014 for q_0, we get u_0 to be approximately 5 and then t_0 is nearly $0.96H_0^{-1}$, that is, nearly equal to the Hubble time.

The deceleration parameter q_0 provides a measure of the slowing down of the expansion of the universe. This dimensionless parameter can, of course, be defined for any time t, and in that case it could be called the deceleration function $q(t)$:

$$q(t) = -\ddot{R}(t)R(t)/\dot{R}^2(t). \tag{3.32}$$

Convenient expressions can be found for $q(t)$ in terms of the parameters θ and u introduced above in the cases $k = 1$ and $k = -1$ respectively. Consider the case $k = 1$. In this case, with the use of (3.1) (with $p = 0$) and (3.2c) we get

$$q(t) = (4\pi G/3)\varepsilon R^2/[-c^4 + (8\pi G/3)\varepsilon R^2]. \tag{3.33}$$

Evaluating (3.33) at $t = t_0$ we get q_0 in terms of ε_0 and R_0, whence we get

$$(2q_0 - 1)/q_0 R_0 = 3c^4/4\pi G\varepsilon_0 R_0^3. \tag{3.34}$$

Equations (3.19) and (3.34) imply

$$(4\pi G/3c^4)\varepsilon_0 R_0^3 R^{-1} = (1 - \cos\theta)^{-1}. \tag{3.35}$$

Eliminating ε from (3.33) with the use of (3.15) and using (3.35), we get

$$q(t) = (1 + \cos\theta)^{-1}. \tag{3.36}$$

Thus as θ varies from 0 to 2π during one cycle, $q(t)$ rises from $\frac{1}{2}$ to infinity and then drops to $\frac{1}{2}$ again. An analysis similar to the one above yields for the case $k = -1$ the following expression for q:

$$q(t) = (1 + \cosh u)^{-1}. \tag{3.37}$$

Thus as u varies from 0 to infinity, q decreases steadily from $\frac{1}{2}$ to zero. In the case $k = 0$, we find with the use of (3.1) (with $p = 0$) and (3.2b), that $q(t)$ remains constant, at the value $q = \frac{1}{2}$.

3.3 Solution for pure radiation

When the cosmological fluid is dominated by radiation, as was presumably the case in the early universe, the equation of state can be taken as

$$p = \tfrac{1}{3}\varepsilon. \tag{3.38}$$

In this case (3.1) reduces to the following equation

$$\ddot{R}c^2 = -(8\pi G/3)\varepsilon R. \tag{3.39}$$

Equation (3.36) can now be integrated to give the following relation:

$$\varepsilon/\varepsilon_0 = (R_0/R)^4. \tag{3.40}$$

The relation corresponding to (3.13) can be written in this case as follows:

$$c^2 k/R_0^2 = (q_0 - 1)H_0^2, \tag{3.41}$$

while the relation corresponding to (3.16) is as follows:

$$(\dot{R}/R_0)^2 = H_0^2(1 - q_0 + q_0 R_0^2/R^2). \tag{3.42}$$

Equation (3.42) can be expressed as an integral as follows:

$$t = H_0^{-1} \int_0^{R/R_0} (1 - q_0 + q_0/x^2)^{-1/2} \, dx, \tag{3.43}$$

with $t = 0$ being the value of t for which $R(t) = 0$. Explicit solutions can be obtained as before, but they are not of much physical interest as the present universe is far from radiation dominated. The behaviour near $t = 0$ is interesting and is considered below. One point worth noting is that in the case $k = 0$ the deceleration function $q(t)$ is constant at $q = 1$, as can be readily verified with the use of (3.2b) and (3.39).

3.4   Behaviour near $t = 0$

It is of considerable interest to determine the behaviour of the function $R(t)$ near the beginning of the universe, that is, near $t = 0$. This behaviour will be used later when we study the early universe. Consider first the zero pressure case. In this case ε varies as R^{-3} so that εR^2 varies as R^{-1}. Thus in all three cases (3.2a)–(3.2c) near $t = 0$ the following relation holds:

$$\dot{R}^2 = 2u\varepsilon_0 R_0^3 R^{-1}, \quad u \equiv 4\pi G/3c^2. \tag{3.44}$$

In the case $k = 0$ this equation holds exactly (for zero pressure). Equation (3.44) can be integrated readily to give the following behaviour for R:

$$R(t) = (\tfrac{3}{2})^{2/3} (2u\varepsilon_0)^{1/3} R_0 t^{2/3}, \tag{3.45}$$

so that R varies as $t^{2/3}$ near $t = 0$, for zero pressure and all three values of k.

Consider next the pure radiation cases given by $p = \tfrac{1}{3}\varepsilon$. In this case ε varies as R^{-4} (see (3.40)), so that εR^2 varies as R^{-2}. Thus in this case too the first terms in (3.2a) and (3.2c) can be ignored near $t = 0$, and all three equations can be written as follows (with the use of (3.40)):

$$\dot{R}^2 = 2u\varepsilon_0 R_0^4 R^{-2}. \tag{3.46}$$

Again in the case $k = 0$ this equation holds exactly. This equation can be integrated to yield the following behaviour for R:

$$R(t) = (8u\varepsilon_0)^{1/4} R_0 t^{1/2}. \tag{3.47}$$

3.5   Exact solution connecting radiation and matter eras

More general equations of state than the cases of zero pressure and pure radiation mentioned above have been considered by Chernin (1965, 1968),

McIntosh (1968) and Landsberg and Park (1975). In this section we give an exact solution for an equation of state which is such that for small values of R it approximates to that of pure radiation, that is, $p = \frac{1}{3}\varepsilon$, while for large values of R the ratio between the pressure and density behaves like R^{-2}, that is, the pressure becomes negligible. We make the ansatz that the mass-energy density is given as a function of R as follows:

$$\varepsilon = AR^{-4}(R^2 + b)^{1/2}, \tag{3.48}$$

where A, b are positive constants. We see that (3.48) implies that for small R, the function ε behaves like R^{-4}, while for large R it behaves like R^{-3}. These are indeed the cases of pure radiation and zero pressure, given respectively by (3.40) and (3.15). We note further that when $b = 0$, (3.48) reduces to the zero pressure case given by (3.15).

We combine Equations (3.2a)–(3.2c) into the following one,

$$\dot{R}^2 = -kc^2 + 2u\varepsilon R^2, \tag{3.49}$$

(with u given by (3.44)) and substitute for ε from (3.48) to get the following equation:

$$\dot{R}^2 = -kc^2 + 2uAR^{-2}(R^2 + b)^{1/2}. \tag{3.50}$$

This equation can be expressed as the following integral:

$$t = \int_0^R [-kR^2c^2 + 2uA(R^2 + b)^{1/2}]^{-1/2}R \, dR. \tag{3.51}$$

This integral can be simplified by the following substitution:

$$x = (R^2 + b)^{1/2}, \tag{3.52}$$

which transforms (3.51) as follows:

$$t = \int_{b^{1/2}}^{(R^2 + b)^{1/2}} (-kx^2c^2 + 2uAx + c^2kb)^{-1/2}x \, dx. \tag{3.53}$$

Consider the three cases $k = 1, 0, -1$ separately.

(i) Case $k = 1$.

In this case (3.53) can be integrated to yield the following parametric relation between R and t:

$$(R^2 + b)^{1/2} = \frac{uA}{c^2} + \frac{v}{c}\sin\theta, \quad v = \left(bc^2 + \frac{u^2A^2}{c^2}\right)^{1/2}, \tag{3.54a}$$

$$tc^3 = uA(\theta - \theta_0) - cv(\cos\theta - \cos\theta_0), \tag{3.54b}$$

where θ_0 is the value of θ for which R vanishes, that is,

$$b^{1/2}c^2 = uA + cv \sin \theta_0. \tag{3.54c}$$

(ii)   Case $k = 0$.

In this case R and t are related as follows:

$$R^2 = -b + (wt + b^{3/4})^{4/3}, \quad w \equiv \tfrac{3}{2}(2uA)^{1/2}. \tag{3.55}$$

(iii)   Case $k = -1$.

In this case (3.53) can be integrated to give the following parametric relation between R and t:

$$(R^2 + b)^{1/2} = -\frac{uA}{c^2} + \frac{v}{c}\cosh \psi, \tag{3.56a}$$

$$tc^3 = -uA(\psi - \psi_0) + cv(\sinh \psi - \sinh \psi_0), \tag{3.56b}$$

where ψ_0 is the value of ψ for which R vanishes, that is,

$$b^{1/2}c^2 = -uA + cv \cosh \psi_0. \tag{3.56c}$$

To find the pressure, we first take the derivative of (3.50) with respect to t and cancel a factor $\dot{R}$ to get the following expression for $\ddot{R}$:

$$\ddot{R} = -uAR^{-3}(R^2 + 2b)(R^2 + b)^{-1/2}. \tag{3.57}$$

From (3.1) we get p as follows:

$$3p = -(uR)^{-1}\ddot{R} - \varepsilon, \tag{3.58}$$

so that, with the use of (3.57), we arrive at the following expression for p:

$$p = (bA/3)R^{-4}(R^2 + b)^{-1/2}. \tag{3.59}$$

The equation of state is given parametrically by (3.48) and (3.59). The condition that as R tends to zero the relation between p and ε tends to $p = \tfrac{1}{3}\varepsilon$ is automatically satisfied by ε and p given by (3.48) and (3.59) respectively.

We get the following value for the ratio of the pressure and the mass-energy density:

$$p/\varepsilon = (b/3)(R^2 + b)^{-1}. \tag{3.60}$$

Thus near $R = 0$ this ratio is $\tfrac{1}{3}$ while as R tends to large values the ratio behaves as R^{-2}, that is, the pressure becomes negligible compared to the mass-energy density, as is indicated by observations.

In the case $k = 1$, we have $R = 0$ at $t = 0$ (for $\theta = \theta_0$), and the maximum value of R occurs at $\theta = \pi/2$, at the value of t given by

$$tc^3 = Tc^3 = uA(\pi/2 - \theta_0) + cv \cos \theta_0, \qquad (3.61)$$

the corresponding value of R being given by the following expression:

$$R = \left(\frac{2uA}{c^3}\right)^{1/2}\left(v + \frac{uA}{c}\right)^{1/2}. \qquad (3.62)$$

After the maximum, R decreases steadily to zero in the manner of a cycloid considered earlier. This case can be considered as a generalized cycloid, the whole cycle lasting for a period of $2T$, the final value of θ being $\pi - \theta_0$. The behaviour of $R(t)$ is thus very similar to the case of pure radiation or zero pressure for $k = 1$.

In all three cases $R(t)$ behaves as $t^{1/2}$ for small t, which is consistent with (3.47). In the case $k = 0$ it is readily seen that for large t, $R(t)$ behaves like t. In the case $k = -1$, large values of R and t occur for large values of the parameter ψ, and for such values both Rc and tc^2 behave like $v\, e^\psi$, so that $R(t)$ tends to infinity like ct, in the manner of the zero pressure case with $k = -1$.

It is of some interest to note that the deceleration function, defined by (3.32), is given by the following expression for this solution:

$$q(t) = uA(R^2 + 2b)/\{(R^2 + b)^{1/2}[-kR^2c^2 + 2uA(R^2 + b)^{1/2}]\}. \quad (3.63)$$

The deceleration function takes the following simple form for the case $k = 0$:

$$q(t) = \tfrac{1}{2}(R^2 + 2b)/(R^2 + b). \qquad (3.64)$$

This function tends to unity as R tends to zero, which is consistent with the fact that for the case of pure radiation and $k = 0$, the deceleration function remains constant at the value $q = 1$ (see the end of Section 3.3). As R tends to infinity, $q(t)$ tends to $\tfrac{1}{2}$, consistent with the zero pressure, $k = 0$ case (see the end of Section 3.2).

3.6   The red-shift versus distance relation

In Section 2.4 we considered the relation between the red-shift and distance for small values of r, $t - t_0$, l, etc. (see (2.97), (2.99), (2.100) and (2.102)). In this section we want to extend that analysis to arbitrary values of the red-shift, etc., with the use of the exact solution for zero pressure.

Let a light ray emitted at $t = t_1$ from the position $r = r_1$ radially be received at the position $r = 0$ at time $t = t_0$. Denoting by R_1 the value of R at t_1, the red-shift z is given as follows (see (2.85)):

$$1 + z = R_0/R_1. \tag{3.65}$$

We consider the analogue of (2.86) for $k \neq 0$ to get the following equation:

$$\int_0^{r_1} (1 - kr^2)^{-1/2} \, dr = c \int_{t_1}^{t_0} \frac{dt}{R(t)} = c \int_{R_1}^{R_0} \frac{dR}{R\dot{R}}. \tag{3.66}$$

We now substitute for $\dot{R}$ from the exact solution for zero pressure given by (3.16), and transform to the integration variable $x = R/R_0$, to get

$$\int_0^{r_1} (1 - kr^2)^{-1/2} \, dr = c(R_0 H_0)^{-1} \int_{(1+z)^{-1}}^1 (1 - 2q_0 + 2q_0/x)^{-1/2} x^{-1} \, dx. \tag{3.67}$$

It can be shown that for all three values of k, the expression for r_1 is the same, as follows:

$$r_1 = c\{zq_0 + (q_0 - 1)[-1 + (2q_0z + 1)^{1/2}]\}/[H_0 R_0 q_0^2 (1 + z)]. \tag{3.68}$$

For large values of the red-shift z it is convenient to define a *luminosity distance*, measured by comparison of *apparent luminosity* and *absolute luminosity*, which are respectively the radiation received by an observer per unit area per unit time from the source, and the radiation emitted by the source per unit solid angle per unit time. The luminosity distance, d_L, is given as follows (see, for example, Weinberg (1972, p. 421)):

$$d_L = r_1 R_0^2/R_1. \tag{3.69}$$

With the use of (3.65) and (3.68), this can be written as follows:

$$d_L = R_0 r_1 (1 + z) = c(H_0 q_0^2)^{-1}\{zq_0 + (q_0 - 1)[-1 + (2q_0z + 1)^{1/2}]\}. \tag{3.70}$$

For small values of z we get

$$d_L = cH_0^{-1}[z + \tfrac{1}{2}(1 - q_0)z^2]. \tag{3.71}$$

This equation is independent of models and can be derived using kinematics only, like (2.102).

3.7   Particle and event horizons

In Section 3.2 we obtained exact and explicit solutions for $R(t)$ for zero pressure, that is, in the matter-dominated era. This solution can be used to illustrate certain limitations of our vision of the universe first pointed out by Rindler (1956). Here we follow closely the discussion of this question given by Weinberg (1972, p. 489). Consider an observer situated at $r = 0$. Let another observer situated at $r = r_1$ emit a light signal at time t_1. Suppose this light signal reaches the first observer at time t. Assuming light to be the fastest of any signals, the only other signals emitted at time t_1 that the first observer receives by time t are from radial coordinates $r < r_1$. Extending (2.86) to the two non-zero values of k, we see that r_1 is determined as follows:

$$\int_0^{r_1} dr/(1 - kr^2)^{1/2} = c \int_{t_1}^{t} dt'/R(t'). \tag{3.72}$$

If the t' integral in (3.72) diverges as t_1 tends to zero, then r_1 can be made as large as we please by taking t_1 to be sufficiently small. Thus in this case in principle it is possible to receive signals emitted at sufficiently early times from any comoving particle, such as a typical galaxy. If, however, the t' integral converges as t_1 tends to zero, then r_1 can never exceed a certain value for a given t. In this case our vision of the universe is limited by what Rindler has called a *particle horizon*. It is possible to receive signals at time t from comoving particles that are within the radial coordinate r_h, which is a function of t, given as follows:

$$\int_0^{r_h} dr/(1 - kr^2)^{1/2} = c \int_0^{t} dt'/R(t'). \tag{3.73}$$

The proper distance d_h of this horizon is

$$d_h(t) = R(t) \int_0^{r^h} dr/(1 - kr^2)^{1/2} = cR(t) \int_0^{t} dt'/R(t'). \tag{3.74}$$

From (2.109a) we see that if the mass-energy density ε varies as $R^{-2-\delta}$ for some positive δ, as R goes to zero, the k on the left hand side of this equation can be neglected and it is readily seen that $R(t)$ behaves as $t^{2/(2+\delta)}$. In this case the t' integral in (3.73) converges as t_1 goes to zero and a particle horizon is present. This is the case in the solution for zero pressure considered in Section 3.2. If the largest contribution to the t' integral comes from the matter-dominated era, we can use (3.17) to express d_h as follows:

$$d_h(t) = \begin{cases} cR_0^{-1}H_0^{-1}(2q_0-1)^{-1/2}R(t)\cos^{-1}[1-q_0^{-1}R_0^{-1}(2q_0-1)R(t)], \\ \qquad\qquad\qquad\qquad\qquad\qquad\qquad q_0 > \tfrac{1}{2} \quad (k=1), \\ 2cH_0^{-1}[R(t)/R_0]^{3/2}, \qquad\qquad\quad q_0 = \tfrac{1}{2} \quad (k=0), \\ cR_0^{-1}H_0^{-1}(1-2q_0)^{-1/2}R(t)\cosh^{-1}[1-q_0^{-1}R_0^{-1}(1-2q_0)R(t)], \\ \qquad\qquad\qquad\qquad\qquad\qquad\qquad q_0 < \tfrac{1}{2} \quad (k=-1). \end{cases}$$

$$(3.75)$$

It can be shown that in the limit of small t, for the early epoch of the matter-dominated era, one gets the following expression for d_h:

$$d_h(t) \rightarrow cH_0^{-1}(q_0/2)^{-1/2}(R/R_0)^{3/2} \approx ct/3. \qquad (3.76)$$

Here $R(t)$ is much smaller than R_0. From (3.75) it is clear that for $q_0 \leqslant \tfrac{1}{2}$, $R(t)$ increases without limit as t tends to infinity, so that $d_h(t)$ increases faster than $R(t)$ and the particle horizon will eventually include all comoving particles, given sufficient time. For $q_0 > \tfrac{1}{2}$, the universe is spatially finite, with a circumference given by

$$L(t) = 2\pi R(t). \qquad (3.77)$$

(See the discussion following (2.58).) At any time t we can see a fraction of this circumference given by (3.13) and (3.75) as follows:

$$d_h(t)/L(t) = (2\pi)^{-1}\cos^{-1}[1-q_0^{-1}R_0^{-1}(2q_0-1)R(t)]. \qquad (3.78)$$

Comoving particles within this fraction are visible. When $R(t)$ reaches its maximum value given by (3.21), this fraction will be $\tfrac{1}{2}$, and we shall see all the way to the 'antipodes'. This fraction remains less than unity until $R(t)$ reaches zero again, so we shall not be able to see all the way around the universe until that happens. If $q_0 = 1$ and $H_0^{-1} = 13 \times 10^9$ years, the present circumference is 82×10^9 light years and the particle horizon is at 20×10^9 light years.

There may be some events in some cosmological models that we shall never see. It is clear from (3.72) that an event that occurs at time t_1 at the coordinate value r_1 will become visible at $r = 0$ at a time t given by (3.72). If the t' integral diverges as t tends to infinity (or at the time that R reaches zero again), then it will be possible to receive signals from any event. However, if the t' integral converges for large t then we can receive signals from only those events for which

$$\int_0^{r_1} dr/(1-kr^2)^{1/2} \leqslant c \int_{t_1}^{t_{max}} dt'/R(t'), \qquad (3.79)$$

where t_{max} is either infinity or the time of the next contraction: $R(t_{max}) = 0$. This is referred to by Rindler as an *event horizon*. It is readily verified that for $q_0 < \frac{1}{2}$ or $q_0 = \frac{1}{2}$, the t' integral diverges as t tends to infinity so that there is no event horizon. For $q_0 > \frac{1}{2}$, $t_{max} = 2T$, where T is given by (3.21). In this case an event horizon exists and the only events occurring at time t_1 that will be visible before R reaches zero again are those within a proper distance $d_E(t_1)$ given as follows:

$$d_E(t_1) = cR(t_1) \int_{t_1}^{t_{max}} dt'/R(t')$$

$$= cR_0^{-1}H_0^{-1}(2q_0 - 1)^{-1/2}R(t_1)\{2\pi - \cos^{-1}[1 - q_0^{-1}R_0^{-1}(2q_0 - 1)R(t_1)]\}.$$

$$(3.81)$$

If $q_0 = 1$ and $H_0^{-1} = 13 \times 10^9$ years, then the only events occurring now that will ever become visible are those within a proper distance of 61×10^9 light years.

4

The Hubble constant and the deceleration parameter

4.1 Introduction

In the last two chapters we developed the mathematical framework, both kinematical and dynamical, to study various cosmological models that may represent, albeit as an idealization, the universe that we inhabit. In this chapter we discuss in some detail the observational aspects that must be considered to connect the models to reality. We will first give an account of earlier developments of this subject and mention more recent work towards the end of Section 4.4.

Two of the most important observational parameters in cosmology are the present values of Hubble's constant H_0 and the deceleration parameter q_0. Hubble's constant determines the present rate of expansion of the universe through the first term on the right hand side of (2.102). We write the following approximate form of this equation here again for convenience:

$$z = H_0 l/c + \tfrac{1}{2}(1 + q_0)H_0^2 l^2/c^2. \tag{4.1}$$

Thus in the limit of small distances the red-shift is given by H_0 times the distance divided by c. The deceleration parameter determines the rate at which the expansion is slowing down (or, in the unlikely case, speeding up). As we see in (4.1), q_0 occurs in the second order term in a power series expansion in terms of l, the distance, Thus q_0 is determined by galaxies which are further than the ones from which H_0 is determined.

As we saw in the last chapter in the case of the Friedmann models, the parameters H_0 and q_0 determine these models completely. For example, if no pressure exists, the age of the universe t_0 is given in terms of H_0, q_0 by (3.18). We then get three possibilities. In the cases $k = 1, 0, -1$ we get $q_0 > \tfrac{1}{2}$, $q_0 = \tfrac{1}{2}$ and $q_0 < \tfrac{1}{2}$ respectively. In these cases the age of the universe is given respectively by (3.23), (3.25) and (3.31). Thus if we knew all three quantities H_0, q_0, t_0 precisely, we could, in principle, decide which of the three models is correct, assuming, of course, zero pressure and other implicit assumptions that go into the definition of these models. We could then know all the large-scale physical properties of the universe.

Although, in principle, the determination of H_0 and q_0 is straightforward, in practice many difficulties arise and in this chapter we will consider some of these difficulties. This chapter is based mainly on the reviews by Sandage (1970, 1987), Gunn (1978) and Longair (1978, 1983).

4.2   Measurement of H_0

As mentioned earlier H_0 is measured from 'local' galaxies which are relatively nearby, whereas q_0 requires consideration of more distant galaxies. The first complication in measuring H_0 is that galaxies possess random motion of the order of 200 km s^{-1} which is caused by local gravitational perturbation, or 'lumpiness' of the galactic distribution, on a scale of about two million light years, which is the size of a small cluster of galaxies. For a large cluster which has rotational velocity about some centre, this random motion can be much higher. One can take account of this random motion, but for this one has to take a large sample. Secondly, there may be local anisotropy, but on quite a large scale, which may distort the velocity field in some directions for red-shifts which imply velocities smaller than about 4000 km s^{-1}. This anisotropy may arise partly due to an abnormal concentration of groups of galaxies such as the Virgo cluster on a scale of about 30 million light years.

Another complication in the measurement of H_0 is the rotational motion of the Sun about the galactic centre of the Milky Way, which amounts to approximately 300 km s^{-1} in the direction of Cygnus. This velocity is an appreciable fraction of the recessional velocities of nearby galaxies in the direction of Cygnus, so this effect appears as an added anisotropy in the observed velocity field. To map this velocity field precisely, accurately subtracting any spurious velocities, requires data from the Southern Hemisphere, which have only recently been forthcoming.

Thus an accurate measurement of H_0 requires precise distance determinations of nearby objects. Distance calibration is a stepwise process in which errors proliferate at each step. First one measures the apparent brightness of well-known objects in nearby galaxies, which can be resolved optically. If the absolute brightnesses of these objects are known from another source, the distance can be determined by the inverse-square law of the falling of the intensity. Because the absolute luminosities can be related to the periods of Cepheid variable stars, these stars are excellent indicators of distance.

The term Cepheid variables derives from a particular member of this class known as Delta Cephei. In the early part of this century H. S. Leavitt and H. Shapley found a relationship between the observed period of variation of the Cepheids and their intrinsic brightness. In 1923 Hubble was for the first time able to resolve the nearby galaxy Andromeda into separate stars, and locate Cepheid variables in it. Using the Leavitt–Shapley relation he concluded that the Andromeda nebula was at a distance of 900 000 light years, which was clearly outside our galaxy, since it was more than ten times further than the most distant object known in our galaxy. Later, however, Baade (1952) and others showed that there are, in fact, two types of Cepheid variables, and that those that Leavitt and Shapley observed and those that Hubble observed belong to the two different types, so that Hubble used the wrong period–luminosity relation. The distance to the Andromeda nebula turns out to be over two million light years. (See Figs 4.1 and 4.2 for further information about Cepheid variables.)

The distance range over which H_0 can be determined is not very large. It starts from about 10^7 light years, which is far enough so that recessional velocities begin to dominate the random velocities, and ends at about 6×10^7 light years, which is the upper limit for the distance indicators to be resolved by powerful optical telescopes. There are various possible distance indicators in this range, such as red and blue supergiants, the angular size of HII regions, normal novae and possibly supernovae. The nearer of these are first calibrated with Cepheid variables and then used in turn as more distant indicators.

Because of Hubble's error alluded to above, the value of H_0 for more than a decade following 1936 was taken to be about 165 km s^{-1} per million light years or about $538 \text{ km s}^{-1} \text{ Mpc}^{-1}$. In the simplest cosmological models this meant an age of the universe of only 1.8×10^9 years. Even in the 1930s this was known to violate the age of the Earth as known from geological studies, such as the age of the Earth's crustal rocks and the lower limit of 7×10^9 years for the age of the Earth's radioactive elements.

There was a controversy in the 1930s and 1940s as to whether the value of H_0 was wrong, or the Friedmann models. Lemaitre and Eddington, for example, devised models with a 'cosmological constant' (about which we will learn more in the next chapter) to fit the high value of H_0. The controversy was finally settled in the 1950s following the work of Baade cited earlier, which started a detailed recalibration of the period–luminosity relation of Cepheid variables, to which contributions were made by Kraft (1961), Sandage and Tammann (1968, 1969) and others. High-precision photometric methods developed in the 1950s by Eggen, Johnson and others also contributed to this progress. These improved calibrations, and the precise distance determinations in the crucial range for H_0 mentioned earlier, such as some highly resolved systems centred on the giant spiral M81, have considerably improved the measurement of H_0. There is, however, still an uncertainty, the present range of values being $15 \leqslant H_0 \leqslant 30$ km s^{-1} 10^{-6} (light year)$^{-1}$ or about $50 \leqslant H_0 \leqslant 100$ km s^{-1} Mpc^{-1}. This makes the age of the universe from 13 to 20 billion years approximately. Among those who have contributed to this new determination of H_0 are Sandage and Tammann (1975), de Vaucouleurs (1977), Tully and Fisher (1977) and Van den Bergh (1975).

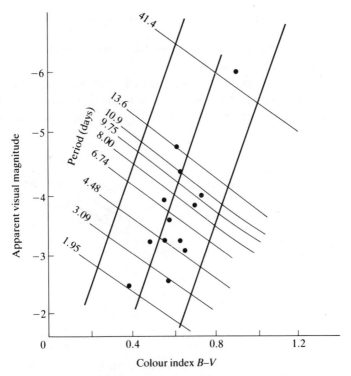

Fig. 4.1. In this diagram the visual luminosity M_V of a star is plotted against its colour B–V which is a measure of its temperature. Here B–V $= 2.5 \log(l_V/l_B) -$ constant, where l_V, l_B are the luminosities integrated over the visual and blue ranges of the spectrum, respectively. Pulsating stars lie in the region between the two outer lines with positive slope. The stars pulsate with periods that increase with increasing luminosity. Cepheids of the same period can differ in absolute luminosity by one magnitude, the bluer Cepheids being brighter.

4.3 Measurement of q_0

If one had knowledge of 'standard candles', that is, objects of fixed, known absolute luminosities, then the apparent luminosities of these objects would be a measure of their distance, and by determining their red-shifts one could plot a graph of red-shift versus apparent luminosity, from which, in principle, one could read off the values of H_0 and q_0. The apparent luminosity of a source is usually described by its so called bolometric magnitude, denoted by m_{bol}. For small red-shifts z the following relation obtains between m_{bol} and z (see the Appendix to this chapter for a definition of m_{bol} and a derivation of this relation, (4.17) below):

$$m_{bol} = 5 \log_{10}(cz) + 1.086(1 - q_0)z + \text{constant}. \tag{4.2}$$

The constant contains H_0 (see Appendix). This relation is true for all Friedmann models, but for small z.

Equation (4.2) is useful because it relates quantities m_{bol} and z which are directly measurable. All observations confirm the leading term in (4.2). Figure 4.3 gives one of these observational plots. The figure has data for

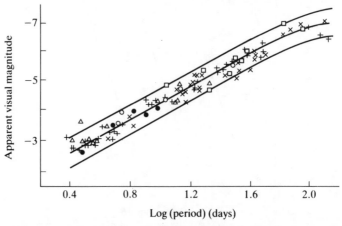

Fig. 4.2. In this diagram the apparent visual magnitude is plotted against the logarithm of the period (in days). The scatter in this diagram is caused by variation of the colour indicated in Fig. 4.1. An accurate calibration can be made if this P–L (period–luminosity) diagram is considered in conjunction with the colour of the Cepheids. Calibrating Cepheids are the galactic cluster Cepheids (solid circles) and the h and Perseus Association (open circles). Other symbols represent Cepheids belonging to the Local Group of galaxies.

42 clusters of galaxies, each of which has a good distance indicator, which is the brightest in the cluster. The small horizontal dispersion about the line, with the theoretical slope of 5, shows the near constancy of absolute luminosity for galaxies chosen this way. It is clear, however, from (4.2) that the value of q_0 cannot be determined from the data in Fig. 4.3, and that one must resort to much higher red-shifts. The data of Fig. 4.3 extend only till $z = 0.46$, and for this kind of red-shift any significant variation in q_0 which could decide between different models gives a variation in m_{bol} which is equivalent to the scatter of galaxies about the mean line, and for this reason not very useful. Further, there are uncertainties in the various corrections to observed magnitudes which are themselves comparable to the variation. One also requires knowledge of the way absolute luminosities evolve during the period which has elapsed since light left those distant galaxies, due to evolution of their stellar content.

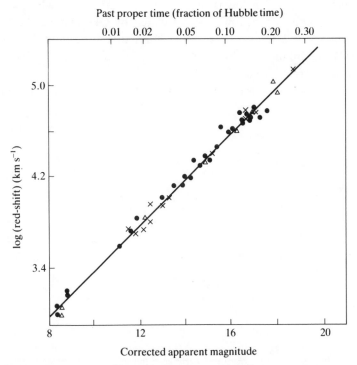

Fig. 4.3. This is a Hubble diagram for 42 galaxies in clusters (see paragraph following Equation (4.2)). Triangles represent non-radio sources measured by W. A. Baum. Crosses represent radio sources and closed circles represent other non-radio sources. These were measured by the 200 in telescope at Mount Palomar. (Sandage, 1970).

Figure 4.4 displays an idealized version of Fig. 4.3, and shows clearly that one needs galaxies with higher red-shifts to distinguish between different values of q_0. The 'test objects' in Fig. 4.3 are giant elliptical galaxies, which tend to be the brightest galaxies in any cluster, and have similar light distribution curves, that is, curves which give a plot of intensity versus wavelength or frequency.

As mentioned earlier, several difficulties arise when one attempts to measure q_0 accurately. One of these is that galaxies do not have well-defined boundaries, so the intensity depends to some extent on the aperture with which one measures it. For this reason all measurements have to be corrected to some adopted 'standard galaxy diameter'.

Every galaxy has an intrinsic frequency distribution of light, that is, an intensity–frequency plot. For distant galaxies this frequency distribution is distorted, because their visual or blue magnitudes reflect their absolute luminosities at higher frequencies than for near galaxies. Thus the left hand side of (4.17) below is replaced by $m - M - k(z)$, where $k(z)$ is an explicitly known function of z, calculated by Oke and Sandage (1968), known as the k term. In the earlier alternative procedure due to Baum (1957), the luminosity distribution is measured directly for each galaxy and no k term is needed.

Our galaxy absorbs a certain amount of radiation coming from objects

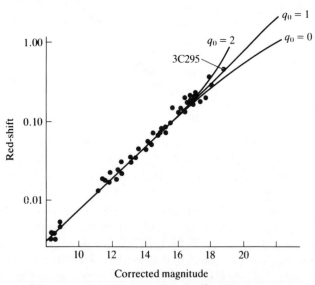

Fig. 4.4. This diagram is an idealized version of Fig. 4.3, showing the extrapolation to regions which would determine the value of q_0, (Weinberg, 1972).

outside the galaxy. Considering the galaxy to be a flat slab, the distance through which light must travel in the galaxy on its way to the observer is proportional to cosec b, where b is the angle between the line of sight and the plane of the galaxy. Due to absorption in the galaxy the light will thus be decreased in intensity by a factor $\exp(-\lambda \csc b)$, where λ is a constant which can be determined from some known extragalactic objects. The distance modulus in (4.17) will then be corrected as follows:

$$(m - M)_{\text{corr}} = m - M - k(z) - A(b), \tag{4.3}$$

where we have approximately, $A(b) = 0.25 \csc b$. This is a somewhat simplified description of the correction due to absorption by the galaxy.

There are still uncertainties in the precise determination of the absolute luminosities of the brightest E galaxies (giant ellipticals). Any change in the estimated distance to nearby objects such as the Hyades or the Virgo cluster would require a corresponding change in these absolute luminosities.

If there is no definite upper limit to the absolute luminosity of a cluster of galaxies, then there would be a tendency to select richer clusters at greater distances resulting in a slight increase of the absolute luminosities of the brightest galaxies with increasing z. This is known as the 'Scott Effect' and may result in a slight overestimation of q_0 but would not have a significant effect on H_0.

The rotation of the Sun about the galactic centre and the existence of a local anisotropy in the galactic velocity field have already been mentioned. The evolutionary effects which were mentioned briefly will be considered in more detail in the Appendix.

The observation and analysis of radio sources have played a significant part in cosmology. For reviews of this topic we refer to Longair (1978, 1983). One of the most important applications of radio astronomy has been the detection and identification of quasars, which are powerful emitters in the radio band. The quasars 3C48 and 3C273, for example, were also identified through optical telescopes and they appeared to be stars, but with peculiar emission lines. These seemed peculiar because the objects were thought to be stars within our galaxy. It was realized later that the emission lines were familiar ones which had been red-shifted by the equivalent of $z = 0.367$ and $z = 0.158$ respectively, so that these were at distances of 5 and 3 billion light years respectively. Many other quasars have since been discovered; the quasar 3C9, for example, has a red-shift of $z = 2.012$. Since the quasars are so bright at such distances, their energy output must be enormous, especially because this energy comes from regions which are only a few light days or weeks across. This follows from the fact that the brightness of the quasars varies substantially over periods of days or weeks. How this enormous energy output is possible from such a small region has been a puzzle for a long time. One of the reasonably successful models is that of a large black hole at the centre of the galaxy which swallows stars, which in the process get disrupted by tidal gravitational forces and give off large amounts of radiation. Such a process can account for the energy output of quasars provided this enormous output does not last for more than a few tens of millions of years at most. There is evidence that quasars do, in fact, only last for a few tens of millions of years.

An intriguing aspect of the quasar problem, which seems worth pursuing carefully, is the fact that there seems to be a cut-off in quasar red-shifts at about $z = 4$. For about ten years the highest quasar red-shift known was $z = 3.53$, although techniques had improved so that higher red-shifts could have been observed. According to M. Smith (see Longair (1983)), there are seven quasars in the Hoag and Smith survey in the red-shift range $2.5 < z < 3.5$, and so eight or nine of them should have been detectable in the Osmer deep survey carried out later, with $3.5 < z < 4.7$, provided the comoving spatial density of quasars remains constant. In fact none was found, although one larger red-shift quasar is now known. However, the question is a statistical one and a great deal more work has to be done before any definite conclusion can be drawn. If indeed there is a cut-off in quasar red-shifts around $z = 4$, the following reasons might be adduced for this phenomenon:

(a) There might be intervening dust in the discs of galaxies so that by the time one gets to distances corresponding to a red-shift of about 3.5, a substantial portion of the celestial sphere might be covered by these discs.

(b) The most prominent emission line through which quasar red-shifts are observed is the Lyman-α line. It is possible that there may be a lack of continuum photons or gas around large red-shift quasars which inhibits the Lyman-α lines.

(c) It is possible that it takes a long time for the black holes, which are at the centre of the largest quasars, to grow. There may be quasars with z much larger than 4, but they may not have grown to the hyperluminous stage.

(d) The dust and gas in the intervening young galaxies may absorb a significant part of the emission from quasars and reradiate in the infrared band.

(e) There is the intriguing possibility that there are no galaxies beyond about $z = 4$, because galaxies may condense out of the intergalactic gas until about $z = 4$.

From the above considerations it would appear that there used to be much more violent activity in the universe at red-shifts of about 2–4 than there is now. This does indicate evolution of the universe and is consistent with the existence of the cosmic background radiation.

Radio astronomy has provided a valuable additional approach to observational cosmology. One of the reasons for its importance is that numerous faint radio sources have been detected, many of which lie presumably at great distances, which have not been optically identified and probably cannot be so identified, at least in the foreseeable future. However, the red-shifts of these sources are for this reason not known, so that one has to follow a programme other than the Hubble programme (outlined above) to elicit information from these faint sources that may be of cosmological interest. Such a programme is that of number counts, in which one determines the number of sources as a function of flux density. It can be shown that in a uniform Euclidean world model, the number of sources N whose flux density is greater than S is proportional to $S^{-3/2}$. By observing and plotting the departure of the actual distribution from this law one can get information about the correct model of the universe. Although there are many uncertainties, some interesting points have emerged. For example, there is evidence that there have been significant variations in the population of radio sources with cosmic epoch. We refer to Weinberg (1972) and Longair (1983) for more details.

4.4 Further remarks about observational cosmology

The distant galaxies that are used for the measurement of q_0 are all in clusters. In this case a substantial proportion of the mass is not in the galaxies, but is distributed smoothly between the galaxies. A galaxy moving through this stuff – whatever form it has – experiences so-called *dynamical friction* (Chandrasekhar, 1960), which is a kind of frictional drag which the moving galaxy experiences by virtue of the high-density gravitational wake behind it. This effect on clusters of galaxies was first studied by Ostriker and Tremaine (1975). The net effect of this is that galaxies which are near the main one are swallowed up by it and its luminosity is thereby increased. The final effect of this on the value of q_0 is somewhat uncertain. Although the brightness of the cannibal galaxy increases, it becomes extended and of low density. Since the luminosity of galaxies is measured in a fixed aperture, it is not clear if the luminosity increases or decreases. The situation is rather complex and a great deal of theoretical and observational work has to be done before this process is fully understood. We refer the reader to Gunn (1978) for further material on this. As is clear from Gunn's article, one of the most important problems is to determine precisely the evolutionary effects on galaxies, clusters of galaxies and quasars.

From (3.14) we recall that if the pressure is negligible, the ratio of the present density to the critical density is twice the deceleration parameter. This ratio is usually denoted by Ω_0 and referred to as the density parameter. The galactic mass density of the universe, that is, the mass of visible matter, is of the order of about a tenth or less of the critical density, so that Ω_0 is around 0.1 or less. However, although there is much uncertainty in the observed value of q_0, indications are that it is about unity or a bit less. There is thus a disagreement between observations and (3.14). This discrepancy has been a long-standing problem in cosmology, and various explanations have been put forward for it. One possibility is that the value of q_0 obtained so far is higher than it should be, because of evolution or selection effects. In fact, it is known that if one determines q_0 solely on the data from quasars, one gets a value somewhat higher than unity. However, if one assumes for the present that q_0 is indeed of order unity, then (3.14) implies that the density of the universe is about $2 \times 10^{-29}\,\text{g cm}^{-3}$. This is an order of magnitude or more higher than that observed. Thus there may be some 'missing mass' which is not directly observable. One possibility is that the missing mass resides in the intergalactic space in clusters of galaxies. If a cluster is gravitationally bound, then by the use of the virial theorem one can estimate its mass, which turns out to be several times higher than would be obtained by adding the masses of individual galaxies (see, for example, Karachentsev (1966)). If this is the case for all or most clusters, the density of the universe would be raised considerably. However, although, for example, the Coma cluster appears to be bound (there is no certainty of this), others like those in Virgo or Hercules are highly irregular and may not be bound.

The missing mass may reside in the space between clusters of galaxies. The total volume outside clusters is approximately 500 times the volume within clusters, so that even a density between clusters which is one-tenth of that within clusters would add significantly to the average density. If indeed there is mass between clusters, presumably it is in a form which does not radiate significantly in the visible spectrum such as atomic or ionized hydrogen, dwarf galaxies which are very faint, black holes, etc. It is uncertain if these or other forms of matter exist in the intergalactic space.

The missing mass may also reside in highly relativistic particles such as cosmic rays, photons, neutrinos or gravitons. These may either be relics from the early universe (see Chapter 7), or may be created in various processes in more recent times. As regards the cosmic background radiation, from the fact that its temperature is 2.7 K and the Stefan–Boltzmann law one can deduce that the associated energy density is about 4.4×10^{-34} g cm^{-3}. The density of cosmic rays and other known forms of radiation is much less than this. As regards 'cosmic background neutrinos', the temperature of these would be $(\frac{4}{11})^{1/3}$ times the temperature of the cosmic background radiation (see Chapter 7 and Weinberg (1977)), or about 2 K. Assuming that the number density of these neutrinos is the same as that of photons, that is, approximately 10^9 for each baryon, this would not make a significant contribution to the overall density if the neutrinos are massless. However, in recent years there have been indications that neutrinos may have a non-zero but small mass, of the order of a few electron volts (recall that the electron has a mass of about half a million electron volts). Thus if neutrinos had a mass of 10 eV, the contribution of neutrinos to the density would be about ten times that due to the visible matter in the universe. However, there are, as usual, many uncertainties in this analysis, and one must wait for more accurate data and theories (see, for example, Tayler (1983)). One point of some interest is that if the density is dominated by massless particles the equation of state becomes that of pure radiation (see Section 3.3) and instead of (3.14) we get $\varepsilon_0/\varepsilon_c = q_0$, and the density required for a given q_0 and H_0 is half of that needed for a zero pressure model.

In the rest of this section we remark on more recent work, following the important review by Sandage (1987). As is clear from the foregoing discussion, one of the most important problems in observational cosmology is the determination of q_0 by comparing (4.2) with observations. Some of the difficulties have already been mentioned in the discussion of Fig. 4.3. Sandage refers to this problem as the '$m(z)$ test'. Following many years of work by various people (Sandage, 1968, 1972a,b, 1975a,b; Sandage and Hardy, 1973; Gunn and Oke, 1975; Kristian, Sandage and Westphal, 1978; Sandage and Tammann, 1983, 1986), Sandage feels that the $m(z)$ test is inconclusive mainly because of uncertainty in the evolution of standard candles. It is clear from (4.2) (see Equation (15) of Sandage (1987)) that for small z, $\log z = 0.2\,m + $ constant. This is indeed indicated by observations until about $z = 0.5$. There are departures from this linear relation between $\log z$ and m for $z > 0.5$. According to Sandage, there may be three different reasons for this departure, as follows: (a) a value of q_0 in the range $0.5 < q_0 < 2$, (b) genuine small departures from linearity for small z, and (c) the combined effects of $q_0 \neq 1$ and evolution of luminosity (see (4.20) below). The reader is referred to studies of the $m(z)$ relation by Lilley and Longair (1984), Lilley, Longair and Allington-Smith (1985) and Spinrad (1986) for more material on this question.

Gross deviations from the $m(z)$ relation – the latter referred to sometimes as the 'Hubble flow' – although sometimes claimed (Arp, 1967, 1980; Burbidge, 1981), have not been substantiated. Large perturbations to the Hubble flow connected with the Local Group of galaxies may, however, exist (see, for example, Davies *et al.*, 1987).

The angular diameter θ of some standard objects also has a dependence on z and q_0 and can be used as a test, as was first suggested by Hoyle (1959). Strictly speaking, the $m(z)$ relation and $\theta(z)$ relation should be derivable from each other, but as θ is directly measurable, this can provide a useful additional check. There are uncertainties in the $\theta(z)$ programme; one has first to give a precise definition of angular diameter (for example, angular size to a given isophote, that is, a contour of equal apparent brightness) that will be valid for sources of all magnitudes and secondly there are the difficulties of evolutionary and selection effects similar to those for the $m(z)$ test (see, for example, Sandage (1972a), Djorgovski and Spinrad (1981)). When discussing the $\theta(z)$ programme Sandage makes an important point about observational cosmology. There are essentially three tests, namely the $m(z)$, $\theta(z)$ and $N(m)$ tests, where $N(m)$ is the number of galaxies brighter than the apparent magnitude m. Sandage thinks that the predictions of the Friedmann models are not confirmed in detail by any of these tests 'using the data as they are directly measured'. To get agreement one usually invokes evolutionary effects with time. This would be justified only if one had independent evidence that the standard model is correct, which is, in fact, the object of the exercise.

An important difficulty is that of selection effects, which, roughly speaking, means that in a sample of sources of limited flux (apparent magnitude), the average absolute luminosity of the nearby members is, in general, less than that of more distant members. Selection effects can cause serious uncertainties, such as in the determination of the value of H_0. The reader is referred to Sandage (1972c), Sandage, Tammann and Yahil (1979), Spaenhauer (1978), Tammann *et al* (1979) and Kraan-Korteweg, Sandage and Tammann (1984) for more material on selection effects, particularly in the form known as the Malmquist bias.

An important observational problem is large scale clustering of galaxies ('superclusters'), first suggested by Hubble (1934). Hubble concluded that the universe was homogeneous on the largest scale that he could measure (to a depth of $m \sim 22$), but that it was clumped or clustered on an intermediate scale. A study by Crane and Saslaw (1986) obtained similar results and drew the same conclusions as Hubble. As regards intermediate structure, an important discovery of the 1980s has been that of 'filaments' along which galaxies tend to concentrate, initially noticed by Peebles and his collaborators (see, for example, Seldner, Siebers, Groth and Peebles (1977)). Unusually large empty regions ('voids') have also been detected. Much work has been done on this matter and is continuing; see for example, studies by Tarenghi *et al* (1979), Gregory, Thompson and Tifft (1981), Kirshner, Oemler, Schechter and Schectman (1981), Gregory and Thompson (1982), Chincarini, Giovanelli and Haynes (1983), Huchra, Davis, Latham and Tonry (1983), and the review by Oort (1983).

The time scale test, one in which one compares the age of the universe from observations and models, has been mentioned earlier. The main uncertainty here is in the value of H_0, which varies by a factor 2. For $H_0 = 50 \text{ km s}^{-1} \text{ Mpc}^{-1}$, the age is about 19.5×10^9 years, whereas for $H_0 = 100 \text{ km s}^{-1} \text{ Mpc}^{-1}$ one gets approximately 9.8×10^9 years. The comparison with observation is somewhat inconclusive (Sandage, Katem and Sandage, 1981; Sandage, 1982).

Sandage suggests the following programme for observational cosmology for the next two decades (writing in 1987). This is a succinct version of the description of the programme given by Sandage (1987).

(a) Proof or otherwise that the red-shift represents a true expansion of the universe.

(b) Proof or otherwise of evolution of galaxies in the look-back time.

(c) Comparison of the value of H_0 with that obtained from the globular cluster time scale. (The globular clusters are among the oldest objects in the Galaxy.)

(d) The compatibility of clustering properties of galaxies with possible variations of the Hubble flow.

(e) Studies of the galaxy luminosity functions for different types of galaxies (see (4.18), (4.19) below).

(f) The detection of $\delta T/T$ fluctuations in the temperature T of the cosmic background radiation at a level of one part in $\sim 10^{5.5}$ on small angular scales. This would have an important bearing on galaxy formation.

Appendix

In this Appendix we derive the formula (4.2), and give some relevant definitions. For more details we refer to Weinberg (1972). The absolute luminosity L of a source is the amount of radiation emitted by the source per unit time. The apparent luminosity l' is the amount of radiation received by the observer per unit time per unit area of the telescopic mirror or plate. In Euclidean space, the apparent luminosity of a source at rest at a distance d would be $L/(4\pi d^2)$, by the usual inverse square law of the decrease of radiation. By analogy with this one defines a *luminosity distance* d_L in the more general case as follows:

$$d_L = (L/4\pi l')^{1/2}. \tag{4.4}$$

By taking into account the red-shift of the moving source one can show that in the general case the apparent luminosity is related to the absolute luminosity as follows:

$$l' = LR^2(t_1)/4\pi R^4(t_0)r_1^2. \tag{4.5}$$

(See Equation (14.4.12) of Weinberg (1972).) Here the source is at the coordinate radius r_1, the times t_0 and t_1 being those of the reception and emission of the radiation. Equation (4.5) is valid for all three values of k. From (4.4) and (4.5) we get

$$d_L = R^2(t_0)r_1/R(t_1). \tag{4.6}$$

By generalizing (2.86) to the two other values of k we get

$$c\int_{t_1}^{t_0} dt/R(t) = \int_0^{r_1} dr/(1 - kr^2)^{1/2} = f(r_1), \tag{4.7}$$

where $f(r_1) = \sin^{-1} r_1, r_1, \sinh^{-1} r_1$ according to whether $k = 1, 0, -1$.
Next we write (2.101), with $t = t_1$:

$$z = (t_0 - t_1)H_0 + (t_0 - t_1)^2(\tfrac{1}{2}q_0 + 1)H_0^2 + \cdots. \tag{4.8}$$

Inverting this power series, we get (see Fig. 4.5)

$$t_0 - t_1 = H_0^{-1}z - H_0^{-1}(1 + \tfrac{1}{2}q_0)z^2 + \cdots. \tag{4.9}$$

With the use of (2.97) or the right hand side of (2.99) in (4.7) we get

$$r_1 + O(r_1^3) = cR_0^{-1}[t_0 - t_1 + \tfrac{1}{2}H_0(t_0 - t_1)^2 + \cdots]. \tag{4.10}$$

With the use of (4.9) and (4.10) we get r_1 in terms of the red-shift:

$$r_1 = c(R_0H_0)^{-1}[z - \tfrac{1}{2}(1 + q_0)z^2 + \cdots]. \tag{4.11}$$

Equation (4.6) then gives d_L as a power series in z:

$$d_L = cH_0^{-1}[z + \tfrac{1}{2}(1 - q_0)z^2 + \cdots]. \tag{4.12}$$

This can be transformed to a formula for the apparent luminosity l':

$$l' = L/(4\pi d_L^2) = c^{-2}(LH_0^2/4\pi z^2)[1 + (q_0 - 1)z + \cdots]. \tag{4.13}$$

The apparent luminosity l' is usually expressed in terms of an *apparent bolometric magnitude* m_{bol}, or simply m, which is defined as follows:

$$l' = 10^{-2m/5} \times 2.52 \times 10^{-5} \text{ erg cm}^{-2} \text{ s}^{-1}. \tag{4.14}$$

The *absolute bolometric magnitude M* is defined as the apparent bolometric magnitude the source would have at a distance of 10 pc:

$$L = 10^{-2M/5} \times 3.02 \times 10^{35} \text{ erg s}^{-1}. \tag{4.15}$$

Thus the *distance modulus $m - M$* can be defined as follows:

$$d_L = 10^{1 + (m - M)/5} \text{ pc}. \tag{4.16}$$

Equations (4.12) and (4.16) can now be combined to give the desired relation between the distance modulus (or the bolometric magnitude) and the red-shift:

$$m - M = 25 - 5 \log_{10} H_0 \text{ (km s}^{-1} \text{ Mpc}^{-1})$$
$$+ 5 \log_{10}(cz) \text{ (km s}^{-1}) + 1.086(1 - q_0)z + \cdots. \tag{4.17}$$

The apparent magnitudes m_U, m_B, etc., in the ultraviolet, blue, photographic, visual (see Fig. 4.1), and infrared wavelength bands are defined similarly to (4.15) and (4.16) but with different constants chosen so that

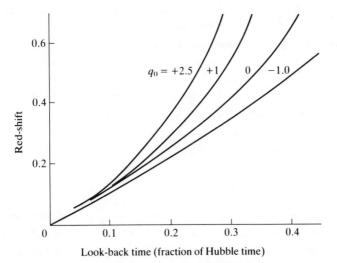

Fig. 4.5. In this diagram Equation (4.9) is illustrated. The 'look-back time' is $t_0 - t_1$ of (4.9) (Sandage, 1970).

all apparent magnitudes will be the same for stars of a certain spectral type and magnitude. The *colour index* is the quantity $m_B - m_V = M_B - M_V$.

We will now give a brief description of the correction to the deceleration parameter due to possible variation of the luminosity L with evolution of galaxies. As we observe distant galaxies, we are looking at earlier times when these galaxies were younger. It is possible that the luminosity of the brightest E galaxies is a function of the time t_1 at which the light was emitted: $L(t_1)$. We see from (4.9) that in this case the L in (4.13) should be replaced by the following expression:

$$L(t_1) = L(t_0)[1 - E_0(t_0 - t_1) + \cdots]$$
$$= L(t_0)[1 - E_0 z/H_0 + \cdots], \tag{4.18}$$

where

$$E_0 = \dot{L}(t_0)/L(t_0). \tag{4.19}$$

Substituting this into (4.13) we readily see that the overall effect is to replace q_0 with q_0^{eff}, where

$$q_0^{\text{eff}} = q_0 - E_0/H_0. \tag{4.20}$$

There are many uncertainties in the value of E_0. Any value of E_0 of the order of $0.04/10^9$ years or above would have a significant effect on the value of q_0^{eff}. It is possible that E_0 is negligible.

We end this Appendix with some remarks about dimensions. It is straightforward to check the dimensions of any of the equations in this book, but the following discussion may help the novice. As usual we denote by L, M, T the dimensions of length, mass and time respectively (the T here is not to be confused with the temperature, which is denoted by T elsewhere in the book). We write $[X]$ for the dimension of the quantity X, and denote by unity the dimension of a dimensionless quantity. The following relations are easy to verify:

$$[G] = M^{-1}L^3 T^{-2}, \quad [c] = LT^{-1}, \tag{4.21a}$$

$$[R] = L, \quad [\dot{R}] = LT^{-1}, \quad [\ddot{R}] = LT^{-2}, \tag{4.21b}$$

$$[z] = [r] = [q_0] = 1, \quad [H_0] = T^{-1}. \tag{4.21c}$$

In (4.21b) and (4.21c) R, r are respectively the scale factor in the Robertson–Walker metric and the coordinate radius. Other similar relations can be derived readily. We will now choose a few equations at random and verify the dimensions of each side of the equations. Consider (3.1), of which the left hand side has dimension LT^{-2} (see (4.21b)). The quantity ε on the right hand side is energy density, that is, energy divided by volume. Since energy has dimension ML^2T^{-2}, we get

$$[\varepsilon] = ML^2T^{-2}/L^3 = ML^{-1}T^{-2}. \tag{4.22}$$

This is the same as the dimension of p, the pressure, which is force per unit area. Force has the dimension MLT^{-2}; dividing this by L^2, the area, yields $ML^{-1}T^{-2}$ as in (4.22). The right hand side of (3.1) thus has dimension

$$[G][\varepsilon][R/c^2] = (M^{-1}L^3T^{-2})(ML^{-1}T^{-2})(L/L^2T^{-2}) = LT^{-2}, \tag{4.23}$$

as required. Consider (3.24), of which the left hand side is clearly dimensionless. Since H_0 has the dimension T^{-1}, clearly $H_0 t$ is also dimensionless. In (3.41) both sides have the dimension T^{-2}. It might be instructive for the reader without much experience of this matter to check in detail the dimension of each equation.

5

Models with a cosmological constant

5.1 Introduction

From (3.1) we see that if we want a *static* solution of Einstein's equations, that is, one in which $\dot{R} = 0$, we must have $\varepsilon + 3p = 0$, which is a somewhat unphysical solution, because, assuming the energy density to be positive, the pressure must be negative. If we demand that the pressure be zero, then the energy density turns out also to be zero.

When Einstein formulated the equations of general relativity in 1915 the expansion of the universe had not been discovered, so that the possibility that the universe may be in a dynamic state did not occur to people. It was natural for Einstein to look for a *static* solution to his cosmological equations. But for the reasons mentioned above such a solution did not appear to exist. Einstein therefore modified his equations by adding the so-called 'cosmological term' to his equation (2.22) , as follows:

$$R_{\mu\nu} - \tfrac{1}{2}g_{\mu\nu}R - \Lambda g_{\mu\nu} = \frac{8\pi G T_{\mu\nu}}{c^4}, \tag{5.1}$$

where Λ is the *cosmological constant*. Equations (2.109a) and (2.109b) are then modified as follows (note that $[\Lambda] = L^{-2}$):

$$3(\dot{R}^2 + c^2 k) = 8\pi G \varepsilon R^2/c^2 + c^2 \Lambda R^2, \tag{5.2a}$$

$$2R\ddot{R} + \dot{R}^2 + kc^2 = -8\pi G p R^2/c^2 + c^2 \Lambda R^2. \tag{5.2b}$$

If we now demand a static solution with $R(t) = R_0$, a constant, and, say, zero pressure, we get the following values:

$$\varepsilon = (c^4 \Lambda/4\pi G), \quad k = \Lambda R_0^2. \tag{5.3}$$

Thus Λ must be positive, and correspondingly, we must choose $k = 1$, so that the universe has positive spatial curvature. This is Einstein's static universe. In later years Einstein regretted adding the cosmological term, because if he had been sure that the universe conformed to his original equations, the fact that no reasonable solutions exist representing a static universe would have led him to infer that the universe is in a dynamic state. He would still not have known if the universe is expanding or contracting, but the discovery of a dynamic state would have been an important one.

Apart from the static solution mentioned above, there are, of course, many dynamic solutions with the cosmological constant. These models were first studied by Lemaitre so they are known as Lemaitre models. In recent years other motivations have been found for introducing a cosmological term and such a term arises in many different contexts. We shall consider some of these later in this chapter and in other chapters. Introducing the cosmological term is like introducing a fictitious 'fluid' with energy momentum tensor $T'_{\mu\nu}$ given by

$$T'_{\mu\nu} = (\varepsilon' + p')u_\mu u_\nu - p'g_{\mu\nu} = (8\pi G/c^4)^{-1}\Lambda g_{\mu\nu}, \tag{5.4}$$

so that the energy density and pressure of this fluid are given by $\varepsilon' = (c^4\Lambda/8\pi G), p' = -(c^4\Lambda/8\pi G)$. For then (5.1) can be written as follows:

$$R_{\mu\nu} - \tfrac{1}{2}g_{\mu\nu}R = (8\pi G/c^4)(T_{\mu\nu} + T'_{\mu\nu}). \tag{5.5}$$

One can follow steps similar to those in Chapter 3 to derive the Lemaitre models. Thus instead of (3.8) we get the following equation:

$$\varepsilon_c = 3c^2 H_0^2/8\pi G = -3kc^4/8\pi GR_0^2 + \varepsilon_0 + c^4\Lambda/8\pi G. \tag{5.6}$$

Recalling the density parameter Ω_0 introduced in the last chapter, (5.6) can be written as follows:

$$c^2k/R_0^2H_0^2 = \Omega_0 - 1 + c^2\Lambda/3H_0^2. \tag{5.7}$$

Equation (3.10) is modified as follows:

$$\varepsilon_0 + 3p_0 = (3/4\pi G)q_0H_0^2c^2 + c^4\Lambda/4\pi G, \tag{5.8}$$

while (3.13) becomes

$$H_0^2(2q_0 - 1) = c^2k/R_0^2 - \Lambda c^2. \tag{5.9}$$

Consider now the solutions that would obtain if we had zero pressure but non-zero Λ. It can be shown after some reduction, in which use is made of (5.7)–(5.9), and the fact that (2.112) and (3.6) remain unaltered, that instead of (3.16) one gets the following equation for $\dot{R}$:

$$\dot{R}^2 = c^2R^{-1}(-kR + \tfrac{1}{3}\Lambda R^3 + \alpha), \tag{5.10}$$

where α is a constant given by $\alpha = R_0^3(H_0^2 c^{-2} - \frac{1}{3}\Lambda + k/R_0^2)$. The behaviour of the solution depends on the pattern of the zeros and turning points of the cubic on the right hand side of (5.10). There are three particular cases of interest, which are dealt with in the following.

(i) de Sitter model

This arises in the case $k = 0$, $\alpha = 0$. With the use of (5.8), (5.9) one can show (in the case $p_0 = 0$), that

$$8\pi G \varepsilon_0 = 3(kc^2/R_0^2 + H_0^2)c^2 - \Lambda c^4. \tag{5.11}$$

Thus if $k = 0$ and $\alpha = 0$, that is, $H_0^2 = \frac{1}{3}\Lambda c^2$, then the mass-energy density also vanishes, and $R(t)$ is proportional to an exponential:

$$R(t) \propto \exp[(\Lambda/3)^{1/2}tc]. \tag{5.12}$$

One gets a similar form for $R(t)$ in the so-called Steady State Theory of Bondi and Gold (1948) and of Hoyle (1948). However, unlike the de Sitter model, which is empty, in the Steady State Theory there is continuous creation of matter due to the so-called C-field.

An interesting property of the metric given by (5.12) is that there is no singularity at a finite time in the past, that is, $R(t)$ does not vanish for any finite value of t (see Fig. 5.1). One can show that this metric has a ten-parameter group of isometries, which is equivalent to 'rotations' in a five-dimensional space with metric whose diagonal elements are $(1, -1, -1, -1, -1)$ and non-diagonal elements zero. This is therefore known as the *de Sitter group*.

(ii) Lemaitre model (Lemaitre, 1927, 1931)

This model corresponds to the solution of (5.10) with $k = 1$ and $\alpha > \alpha_0$, where α_0 is the value of α obtained when Λ has the value in the Einstein

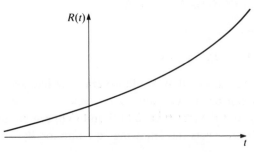

Fig. 5.1. Behaviour of $R(t)$ in the de Sitter model.

static case given by (5.3). From (5.10) we find by differentiation

$$\ddot{R} = -\tfrac{1}{3}c^2\Lambda R - c^2\alpha/2R^2. \tag{5.13}$$

In this model $R(t)$ starts from zero at $t = 0$ and increases at first like $t^{2/3}$. We see from (5.13) that at the initial stage $\ddot{R}$ is negative so the expansion is slowing down. The minimum rate of expansion occurs at $R = (3\alpha/2\Lambda)^{1/3}$, when $\ddot{R} = 0$, after which the expansion speeds up, ultimately reaching the de Sitter behaviour given by (5.12). An interesting property of this solution is that there is a 'coasting period' near the point at which $\dot{R}$ has its minimum, when the value of $R(t)$ remains almost equal to $(3\alpha/2\Lambda)^{1/3}$. By taking $R(t)$ close to this value, we can write an approximate form of the differential equation (5.13) for $k = 1$, as follows:

$$\dot{R}^2/c^2 \simeq -1 + (9\alpha^2\Lambda/4)^{1/3} + \Lambda(R - (3\alpha/2\Lambda)^{1/3})^2, \tag{5.14}$$

which has the following solution:

$$R(t) = (3\alpha/2\Lambda)^{1/3}\{1 + [1 - (9\alpha^2\Lambda/4)^{-1/3}]^{1/2} \sinh(\Lambda^{1/2}(t - t_m)c)\}, \tag{5.15}$$

where t_m is the time at which $\dot{R}$ reaches its minimum. By taking α sufficiently close to $2/(3\Lambda)^{1/2}$, one can make the coasting period arbitrarily long. In the latter half of the 1960s there was some evidence that an excess of quasars with red-shifts approximately equal to 2 might exist. This prompted Petrosian, Salpeter and Szekeres (1967) to invoke the Lemaitre model, because in this model the parameters can be adjusted so that the 'coasting period' causes an excess of quasars with red-shift 2 or so. However, later the statistical evidence for such an excess disappeared. (See Fig. 5.2 for this model.)

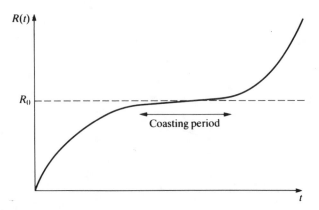

Fig. 5.2. Behaviour of $R(t)$ in the Lemaitre models.

(iii)   Eddington–Lemaitre model

This is a limiting case of the Lemaitre models, which is given this name because it was emphasized by Eddington (1930). In this case $k = 1$ and $\alpha = 2/(3\Lambda)^{1/2}$, which are the values that obtain in the Einstein static model. This model has an infinitely long 'coasting period'. Thus if $R(0) = 0$, then $R(t)$ approaches the Einstein value $(3\alpha/2\Lambda)^{1/3}$ asymptotically from below as t tends to infinity, while if $R(0) = (3\alpha/2\Lambda)^{1/3}$, then $R(t)$ increases monotonically, eventually reaching the de Sitter behaviour as t tends to infinity (see Fig. 5.3). This also shows that the Einstein static model is unstable, at least under perturbations which preserve the Robertson– Walker form of the metric, because when perturbed it will either keep on expanding or keep on contracting.

5.2   Further remarks about the cosmological constant

As is clear from the existence of the Einstein static model, a positive cosmological constant as introduced here represents a repulsive force, so that the attractive force of the matter is balanced by this repulsive force in the Einstein model. In the dynamic models when the galaxies are very far apart after a period of expansion, the attractive force of the matter becomes weak and eventually the repulsive force due to a positive cosmological constant takes over, and one gets asymptotically the de Sitter behaviour. Correspondingly, with a negative cosmological constant one gets an attractive force in addition to the gravitational attractive force already present.

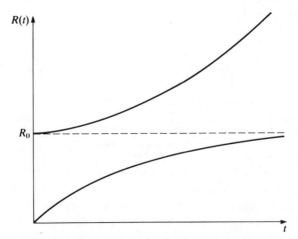

Fig. 5.3. Behaviour of $R(t)$ in the Eddington–Lemaitre model.

We saw in the case of the Friedmann models that a model which expands forever corresponds to $k = 0, -1$, that is, it has infinite spatial volume (the spatial curvature is zero or negative), whereas a model which eventually collapses has $k = 1$, that is, finite spatial volume and positive curvature. This is no longer valid in models with a cosmological constant. We have seen in the case of the Lemaitre models (see Fig. 5.2), that although $k = 1$, the model expands forever. With a negative cosmological constant it is possible to have $k = 0, -1$, and a collapse in the future. This is clear from (5.10), because if Λ is negative eventually the term $\frac{1}{3}\Lambda R^3$ will dominate, so that $R(t)$ cannot be very large, for then $\dot{R}^2$ becomes negative. This will happen regardless of the value of k.

During the mid-seventies there was some evidence that the deceleration parameter q_0 might be negative, that is, the rate of expansion may be increasing. This prompted Gunn and Tinsley (1975) to invoke a Lemaitre model with a positive Λ. However, later considerations of evolutionary effects such as that of galactic cannibalism mentioned in the last chapter modified the value of q_0, so that such an 'accelerating' universe no longer seemed necessary.

The extent to which a cosmological constant is necessary is uncertain. However, to give cosmological studies generality and scope it seems reasonable to consider (H_0, q_0, Λ) as the three unknown parameters of cosmology which have to be determined from observation. The cosmological constant may turn out to be zero, in which case the actual model will be a pure Friedmann one. However, it may also turn out to be non-zero, for while there is no compelling reason for having a cosmological constant, there is also not sufficient reason for its absence. Zel'dovich (1968) has suggested that a term may occur due to quantum fluctuations of the vacuum; in this case the cosmological term becomes a part of the energy–momentum tensor. To consider another motivation for having a Λ term, which arises in the work of Hawking (see citation in Islam (1983b)), we have to know about anti-de-Sitter space, which occurs in a similar manner to the de Sitter space considered above except that Λ is now negative. The metric in this case can be written as (A is a constant with dimension of length)

$$ds^2 = c^2\, dt^2 - A^2 \cos^2 t[d\chi^2 + \sinh^2 \chi(d\theta^2 + \sin^2 \theta\, d\phi^2)]. \quad (5.16)$$

This coordinate system covers only a part of the space. For more details on anti-de-Sitter space the reader is referred to Hawking and Ellis (1973, p. 131). Hawking considers $N = 8$ supergravity theory (see, for example, Freund (1986)) and shows that in this theory a phase transition occurs at a certain critical value of the coupling constant and below this critical value the ground state is an anti-de-Sitter space with a negative cosmological constant. Above the critical value there exists a contribution to the Ricci tensor due to vacuum fluctuations which is equivalent to a positive cosmological constant so that the net effect is that the ground state has an 'apparent' cosmological constant which is zero. Other contexts in which the cosmological constant arises will be mentioned later in this book. Other people have given reasons why Λ is small or zero (Coleman, 1988; Banks, 1988; Morris, Thorne and Yurtsever, 1988). Coleman, for example, suggests quantum tunnelling between separate universes (see Chapter 9 and Schwarzschild (1989)).

It was shown by McCrea and Milne (1934) that many of the properties of the Friedmann and Lemaitre models can be derived from purely Newtonian considerations if one assumes that the universe is in a dynamic state. The cosmological term is introduced by postulating a force which is proportional to the distance between particles (see next section). However, the conceptual basis of this formulation is not sound partly because it does not incorporate the special theory of relativity.

It is clear from the above discussion that it is important to have limits on the cosmological constant. This we will consider in the next section. For a selection of other works on Lemaitre models, we refer to Petrosian and Salpeter (1968), Kardashev (1967), Brecher and Silk (1969), Tinsley (1977), Raychaudhuri (1979), Bondi (1961) (the last three contain reviews).

5.3   Limits on the cosmological constant

From (5.7) and (5.9) we get the following relation:

$$q_0 = \tfrac{1}{2}\Omega_0 - c^2\Lambda/3H_0^2. \tag{5.17}$$

This is the equation which replaces (3.14), the latter being valid for Friedmann models. From (5.17) we get

$$|q_0 - \tfrac{1}{2}\Omega_0| = |c^2\Lambda/3H_0^2|. \tag{5.18}$$

Although the observational values of q_0 and Ω_0 are uncertain, one can reasonably safely say that q_0 lies between -5 and $+5$, and that Ω_0 lies between 0 and 4. The left hand side of (5.18) can then have the maximum value of 7, so that we get

$$|\Lambda| = 21H_0^2/c^2. \tag{5.19}$$

By setting a limit of $100\,\mathrm{km\,s^{-1}\,Mpc^{-1}}$ on H_0, (5.19) leads to a limit of approximately $10^{-54}\,\mathrm{cm^{-2}}$ on the absolute value of Λ (this limit is mentioned by Hawking; see citation in Islam (1983b)).

The above limit comes from cosmological considerations. It is of some interest to see if local considerations can give anything like the same limits. Such a local limit can be obtained by considering the effect of a Λ term on the perihelion shift of Mercury (Islam, 1983b). The Schwarzschild metric is modified as follows by the Λ term (here r has dimension of length):

$$ds^2 = c^2(1 - 2m/r - \tfrac{1}{3}\Lambda r^2)\, dt^2 - (1 - 2m/r - \tfrac{1}{3}\Lambda r^2)^{-1}\, dr^2 - r^2(d\theta^2 + \sin^2\theta\, d\phi^2),$$

$$(5.30)$$

where m is the mass of the Sun, multiplied by G/c^2. It is well known that the usual Schwarzschild solution implies a perihelion shift of Mercury of about 43″ per century. This shift is known with an accuracy of about half a per cent. Using this fact one can show that Λ must satisfy the following inequality (see Islam (1983b) for more details):

$$|\Lambda| < 10^{-42}\ \mathrm{cm}^{-2}. \qquad (5.21)$$

Thus the limit from local considerations is much worse than that derived from cosmology, as expected. One can improve on (5.21) by considering local systems of bigger dimensions, such as the fact that a galaxy is a bound system (Islam and Munshi 1990). For this we consider a typical galaxy such as ours with 10^{11} stars of solar mass, that is, of mass 2×10^{33} g. The matter contained in a disc of diameter 80 thousand light years and thickness 6 thousand light years we imagine to fill a sphere of uniform density with the same average density. The equivalent sphere has a radius of about 19 thousand light years.

Let $\mathbf{r}$ be the position vector of a point with respect to the centre of the spherical galaxy, then we assume the force on a unit mass to be given by (ρ is the density):

$$\mathbf{F} = -\tfrac{4}{3}\pi\rho G\mathbf{r} + \tfrac{1}{3}\Lambda c^2\mathbf{r}. \qquad (5.22)$$

Here the Λ term is the Newtonian form of the cosmological term. As before, a positive Λ implies a repulsive force. The first term on the right hand side of (5.22) represents the usual gravitational force. The galaxy ceases to be a bound system if the right hand side of (5.22) gives an outward force. The condition for this is

$$\Lambda < 4\pi\rho G/c^2. \qquad (5.23)$$

For the dimensions given above we find $4\pi\rho \approx 1.14 \times 10^{-22}$ g cm^{-3}, so that (5.23) gives a limit of approximately 10^{-48} cm^{-2} for Λ. This could also be considered as a limit on the absolute value of Λ, for even if Λ is negative, if its absolute value violated this limit, the effect on the binding of the galaxy would be noticeable. One has to augment this analysis with a general relativistic one by considering the Schwarzschild interior solution, and its modification due to the Λ term. For this and other details we refer to Islam and Munshi (1990).

6

Singularities in cosmology

6.1 Introduction

In Chapter 3 we saw that all the Friedmann models have singularities in the finite past, that is, at a finite time in the past, which we have called $t = 0$; the scale factor $R(t)$ goes to zero and correspondingly some physical variables, such as the energy density, go to infinity. Only exceptionally, such as in the de Sitter or the Steady State models (see Fig. 5.1), is there no singularity in the finite past. But these latter models have some unphysical or unorthodox feature, such as the continuous creation of matter, which is not generally acceptable. The presence of singularities in the universe, where physical variables such as the mass-energy density or the pressure or the strength of the gravitational field go to infinity seems doubtful to many people, who therefore feel uneasy about this kind of prediction of the equations of general relativity. This was partly the motivation with which Einstein searched for a 'unified field theory'. In this connection he says (1950):

The theory is based on a separation of the concepts of the gravitational field and matter. While this may be a valid approximation for weak fields, it may presumably be quite inadequate for very high densities of matter. One may not therefore assume the validity of the equations for very high densities and it is just possible that in a unified theory there would be no such singularity.

There was at one time the feeling that the singularities in the Friedmann models arise because of the highly symmetric and idealized form of the metric, and that, for example, if the metric were not spherically symmetric, the matter coming from different directions might 'miss' each other and not gather at the centre of symmetry, as it does in the (spherically symmetric) Friedmann models. However, it was shown by Penrose and Hawking (1970) that spherical symmetry is not essential for the existence of a singularity. We shall consider this work later.

There are in the main two possible approaches for dealing with the problem of singularities. Firstly, one can try to relax the symmetry conditions inherent in Robertson–Walker metrics and try to determine what the field equations predict in these more general cases. Secondly, one can try to derive some general results about singularities by using reasonable assumptions, say about the energy–momentum tensor, without considering the field equations in detail. The Penrose–Hawking results fall in the latter category. As regards the former approach, the simplest relaxation of the symmetries of the Robertson–Walker metrics (which are homogeneous and isotropic) is to drop the requirement of isotropy and consider metrics which are only homogeneous. A simple example of such a metric was given in (2.48). We shall consider such metrics in some detail in the next section, partly with a view to explaining another approach to the question of singularities, pioneered by Lifshitz and Khalatnikov (1963). There is an extensive literature on singularities and cosmological solutions, incorporating both the approaches mentioned above. This chapter is meant to be only a brief introduction to this work. For more detailed reviews the reader is referred to Hawking and Ellis (1973), Ryan and Shepley (1975), Landau and Lifshitz (1975), MacCallum (1973) and Raychaudhuri (1979).

6.2   Homogeneous cosmologies

In this section we shall derive the metric and field equations for homogeneous (but not isotropic) cosmologies. We shall give the bare essentials here. For more details the reader can consult Landau and Lifshitz (1975, p. 381).

In Section 2.2 we defined a homogeneous space. To continue that discussion, consider the spatial part of the metric (2.34), as follows:

$$dl^2 = h_{ij}(t, x^1, x^2, x^3) \, dx^i \, dx^j, \tag{6.1}$$

where as usual the indices i and j are to be summed over values 1, 2, 3. A metric is homogeneous if after a transformation of the spatial coordinates x^1, x^2, x^3 to new coordinates x'^1, x'^2, x'^3 the metric (6.1) transforms to the following one:

$$dl^2 = h_{ij}(t, x'^1, x'^2, x'^3) \, dx'^i \, dx'^j, \tag{6.2}$$

with the same functional dependence as before of the h_{ij} on the new spatial coordinates. Further, this set of transformations must be able to carry any point to any other point. We saw an explicit example of such a transformation in a simple case in (2.48). One way to characterize the invariance of the metric under spatial transformations is to consider a set of three differential forms $e_m^{(a)} \, dx^m$ (with $a = 1, 2, 3$) which are invariant under these transformations, as follows:

$$e_m^{(a)}(x) \, dx^m = e_m^{(a)}(x') \, dx'^m, \tag{6.3}$$

where we have written x for x^1, x^2, x^3, etc. in the arguments. With the use of these forms a metric invariant under spatial transformations can be constructed as follows (the η_{ab} are six functions of t):

$$dl^2 = \eta_{ab}(e_m^{(a)} \, dx^m)(e_n^{(b)} \, dx^n), \tag{6.4}$$

that is, the three-dimensional metric tensor h_{ij} of (6.2) is given as follows:

$$h_{ij} = \eta_{ab} e_i^{(a)} e_j^{(b)}. \tag{6.5}$$

Note that in (6.3) the $e_m^{(a)}$ on the two sides of the equation are respectively the same functions of the old and new coordinates. We introduce the reciprocal triplet of vectors $e_{(a)}^m$ by the following relations:

$$e_{(a)}^m e_m^{(b)} = \delta_a^b, \quad e_{(a)}^m e_n^{(a)} = \delta_n^m. \tag{6.6}$$

It can be shown after some manipulations (see Landau and Lifshitz (1975, pp. 382–3)), that (6.3) leads to the following equation for the reciprocal triplet $e_{(a)}^m$:

$$e_{(a)}^m \frac{\partial e_{(b)}^n}{\partial x^m} - e_{(b)}^m \frac{\partial e_{(a)}^n}{\partial x^m} = C_{ab}^c e_{(c)}^n, \tag{6.7}$$

where the C_{ab}^c are constants satisfying $C_{ab}^c = -C_{ba}^c$. These are the so-called structure constants of the groups of transformations. If we denote by X_a the following linear differential operator:

$$X_a = e_{(a)}^m \frac{\partial}{\partial x^m}, \tag{6.8}$$

then (6.7) can be written as follows:

$$[X_a, X_b] \equiv X_a X_b - X_b X_a = C_{ab}^c X_c. \tag{6.9}$$

One can now use the Jacobi identity given by

$$[[X_a, X_b], X_c] + [[X_b, X_c], X_a] + [[X_c, X_a], X_b] = 0, \tag{6.10}$$

to derive the following relation for the structure constants:

$$C_{ab}^e C_{ec}^d + C_{bc}^e C_{ea}^d + C_{ca}^e C_{eb}^d = 0. \tag{6.11}$$

The different types of homogeneous spaces correspond to the different inequivalent solutions of (6.11) satisfying the antisymmetry condition

$C^c_{ab} = -C^c_{ba}$. Some solutions are equivalent to each other reflecting the fact that the $e^m_{(a)}$ can still be subjected to a linear transformation with constant coefficients so that the operators X_a are not unique.

There are nine different types of homogeneous spaces that arise from the different inequivalent solutions of (6.11) with the required antisymmetry condition. These are known as the Bianchi types, types I–IX. The Einstein equations for these spaces can be reduced to a system of ordinary differential equations for the $\eta_{ab}(t)$, without the necessity of working out the frame vectors $e^{(a)}_m$, etc. We will consider an application of these results in Section 6.7.

6.3   Some results of general relativistic hydrodynamics

Before considering the results of Penrose and Hawking it is useful to have some idea of relativistic hydrodynamics. The fundamental quantity here is the four-velocity vector u^μ of a continuous distribution of matter in hydrodynamic motion. Thus u^μ is a unit time-like vector. Some of the following formulae are valid for any arbitrary four-vector u^μ. With the use of the covariant derivative $u_{\mu;\nu}$ one can define the following quantities which are of physical significance:

(a)   The scalar expansion $\theta = u^\mu{}_{;\mu}$, which gives the rate at which a volume element orthogonal to the vector u^μ expands or contracts.

(b)   A measure of the departure of the velocity field from geodesic motion is given by the acceleration $\dot{u}_\mu = u_{\mu;\nu}u^\nu$. In the absence of non-gravitational forces, such as in the case of dust (pressure-less matter), the particles follow geodesics and the acceleration vanishes.

(c)   The shear tensor is symmetric, trace-free and is orthogonal to the vector u_μ. It describes the manner in which a volume element orthogonal to u^μ changes its shape, and is given as follows:

$$\sigma_{\mu\nu} = \tfrac{1}{2}(u_{\mu;\nu} + u_{\nu;\mu}) - \tfrac{1}{3}(g_{\mu\nu} - u_\mu u_\nu)\dot\theta - \tfrac{1}{2}(\dot{u}_\mu u_\nu + \dot{u}_\nu u_\mu). \tag{6.12}$$

(d)   A measure of the amount of rotational motion present in the matter is given by the vorticity tensor defined as follows:

$$w_{\mu\nu} = \tfrac{1}{2}(u_{\mu;\nu} - u_{\nu;\mu}) - \tfrac{1}{2}(\dot{u}_\mu u_\nu - \dot{u}_\nu u_\mu). \tag{6.13}$$

One can also define a vorticity vector w^μ as follows:

$$w^\mu = \tfrac{1}{2}\varepsilon^{\mu\nu\rho\sigma}u_\nu u_{\rho;\sigma}, \tag{6.14}$$

where $\varepsilon^{\mu\nu\rho\sigma}$ is the Levi–Civita alternating tensor which is antisymmetric in any pair of indices with $\varepsilon^{0123} = (-g)^{-1/2}$, g being the determinant of the metric. If the vorticity vector or tensor vanishes, the vector u^μ is said to be hypersurface orthogonal and this implies the absence of rotation in some invariant sense (rotation of the local rest frame relative to the compass of inertia; see, for example, Synge (1937), Gödel (1949)).

Next we use (2.12) with u_μ instead of A_μ and make slight changes in the indices to get the following equation:

$$u^\mu{}_{;\alpha;\beta} - u^\mu{}_{;\beta;\alpha} = R^\mu{}_{\nu\beta\alpha}u^\nu. \tag{6.15}$$

In this equation we set μ equal to β and multiply the resulting equation with u^α as follows:

$$u^\alpha(u^\mu{}_{;\alpha;\mu} - u^\mu{}_{;\mu;\alpha}) = R_{\nu\alpha}u^\nu u^\alpha, \tag{6.16}$$

where we have used (2.16). From the Einstein equation (2.22) with (2.23) we readily get

$$R_{\mu\nu} = \frac{8\pi G}{c^4}\left[(\varepsilon + p)u_\mu u_\nu + \tfrac{1}{2}(p - \varepsilon)g_{\mu\nu}\right], \tag{6.17}$$

whence it follows:

$$R_{\mu\nu}u^\mu u^\nu = \frac{4\pi G}{c^4}(\varepsilon + 3p). \tag{6.18}$$

One can use the definitions of expansion, shear, vorticity and acceleration given above to write (6.16) as follows:

$$\theta_{,\alpha}u^\alpha + \tfrac{1}{3}\theta^2 - \dot{u}^\alpha{}_{;\alpha} + 2(\sigma^2 - w^2) = -R_{\mu\nu}u^\mu u^\nu. \tag{6.19}$$

In deriving this relation the following equations have been used (the first one follows by taking the dot-derivative of $u^\mu u_\mu = 1$):

$$\dot{u}_\mu u^\mu = 0, \tag{6.20a}$$

$$\sigma_{\mu\nu}u^\nu = w_{\mu\nu}u^\nu = 0, \tag{6.20b}$$

$$\sigma^2 \equiv \tfrac{1}{2}\sigma_{\mu\nu}\sigma^{\mu\nu}, \tag{6.20c}$$

$$w^2 \equiv \tfrac{1}{2}w_{\mu\nu}w^{\mu\nu}. \tag{6.20d}$$

Equation (6.19) holds for an arbitrary four-vector u^μ. We now let u^μ be the four-velocity of matter, so that (6.18) can be used in (6.19). We then get the following important equation, known as the Raychaudhuri equation (Raychaudhuri, 1955, 1979):

$$\theta_{,\alpha}u^\alpha + \tfrac{1}{3}\theta^2 - \dot{u}^\alpha{}_{;\alpha} + 2(\sigma^2 - w^2) + 4\pi(\varepsilon + 3p)Gc^{-4} = 0. \tag{6.21}$$

The importance of this equation derives from the fact that in one form or another it is used in most if not all singularity theorems of general relativity. To see the relevance of this equation to the question of singularities we consider a simple and somewhat crude analysis. Consider a set of time-like geodesics described by the four-vector u^μ. Let these geodesics be irrotational. Thus we have $\dot{u}^\mu = w = 0$. Let λ be a parameter along a typical geodesic so that $u^\mu = dx^\mu/d\lambda$. Then

$$\theta_{,\alpha} u^\alpha = \frac{\partial \theta}{\partial x^\alpha} \frac{dx^\alpha}{d\lambda} = \frac{d\theta}{d\lambda} = -\tfrac{1}{3}\theta^2 - 2\sigma^2 - 4\pi(\varepsilon + 3p)Gc^{-4}. \qquad (6.22)$$

Now make the assumption that $2\sigma^2 + 4\pi(\varepsilon + 3p)Gc^{-4}$ is greater than a positive constant $\tfrac{1}{3}\xi^2$. Then the behaviour of θ is governed by the following differential equation:

$$d\theta/d\lambda = -\tfrac{1}{3}(\theta^2 + \xi^2), \qquad (6.23)$$

which has the solution

$$\theta = \theta_0 - \xi \tan[(\xi/3)(\lambda - \lambda_0)], \qquad (6.24)$$

θ_0 being the value of θ at $\lambda = \lambda_0$. From this equation it is clear that θ becomes infinite as λ is decreased from the value λ_0 to $\lambda_0 - 3\pi/2\xi$. If, for example, λ denotes the proper time along the geodesic, then this shows that at a finite time in the past the expansion θ becomes infinite. An infinite value of θ indicates that at that point geodesics cross each other and there is a sort of 'explosion' like the big bang. In the Friedmann models u^μ is given by the vector $(1, 0, 0, 0)$ and it is readily verified that θ, which is the covariant divergence of this vector, is given by $3\dot{R}/R$. In the case $k = 0$, for example, from (3.2b) we see that this is proportional to $\varepsilon^{1/2}$. We know that this tends to infinity as the big bang $t = 0$ is approached. Thus the expansion θ tends to infinity at a finite time in the past. The assumption $2\sigma^2 + 4\pi Gc^{-4}(\varepsilon + 3p) = \tfrac{1}{3}\xi^2$ is a limiting case. If $2\sigma^2 + 4\pi Gc^{-4}(\varepsilon + 3p) > \tfrac{1}{3}\xi^2$ the infinity in θ occurs at a shorter distance away from $\lambda = \lambda_0$.

The above somewhat crude analysis can be made more precise, and this is essentially what is done in the singularity theorems. These theorems are very technical and need a great deal of preliminary apparatus. We shall here give only the statement of one of these theorems, but we need some familiarity with singularities.

6.4 Definition of singularities

The question of a definition of singularities in general relativity is a highly complex one and we can only consider a bare outline of the extensive literature on the subject. An excellent account of this topic is given in Hawking and Ellis (1973).

We have encountered a simple case of a singularity in the Friedmann models, where at $t = 0$ the mass-energy density goes to infinity. The mass-energy density is a simple example of the so-called 'curvature scalars' or 'curvature invariants' whose values do not change under a coordinate transformation, so that if they are infinite at a certain point in one coordinate system, they will be infinite at that point in every coordinate system. Another example of a curvature scalar is the Ricci scalar defined by (2.20). It is well known that in empty space (where the Ricci tensor vanishes), there are four curvature invariants, one of these being $R_{\alpha\beta\gamma\delta}R^{\alpha\beta\gamma\delta}$ (see, for example, Weinberg (1972) for a discussion of this). If one of the curvature scalars goes to infinity at a point, that point is a space-time singularity, and cannot be considered as a part of the space-time manifold, whose points are defined to be such that one can introduce a coordinate system so that the metric and its derivatives to second order are well behaved. Such points may be called 'regular' points. However, all the curvature scalars remaining finite at a point does not necessarily imply the point is regular. The usual example of this that is cited is that of the two-dimensional surface of an ordinary cone in three dimensions. The curvature scalars of this surface remain finite as one approaches the apex of the cone, but the latter is not a regular point as it is not possible to introduce any coordinate system that is well behaved at that point. On the other hand, the metric behaving badly at a point does not necessarily mean that the point is singular, because the bad behaviour may be simply due to the unsuitable nature of the coordinate system. These matters are illustrated well by the Schwarzschild metric.

The Schwarzschild solution is given as follows:

$$ds^2 = c^2(1 - 2m/r)\,dt^2 - (1 - 2m/r)^{-1}\,dr^2 - r^2(d\theta^2 + \sin^2\theta\,d\phi^2). \quad (6.25)$$

Here the coefficient of dt^2 goes to infinity at $r = 0$ and that of dr^2 goes to infinity at $r = 2m$. The curvature invariants are well behaved at $r = 2m$, but some of them go to infinity at $r = 0$. Thus the bad behaviour of the metric cannot be removed at $r = 0$, so the latter is a singularity. However, as mentioned earlier, the fact that the curvature invariants are regular at $r = 2m$ does not necessarily mean that the latter is not a singularity. To prove this one would have to find a coordinate system which is well behaved at the point. For a long time after the Schwarzschild solution was discovered, in 1916, such a coordinate system could not be found. It was observed that the radial time-like and null geodesics displayed no unusual behaviour at $r = 2m$. Finally in 1960 Kruskal found the following transformation from (r, t) to new coordinates (u, v) which shows that the point $r = 2m$ is regular:

$$u^2 - v^2 = (2m)^{-1}(r - 2m)\exp(r/2m), \quad v = u\tanh(ct/4m), \quad (6.26)$$

with the metric (6.25) given as follows:

$$ds^2 = r^{-1}(32m^3)\exp(-r/2m)(du^2 - dv^2) - r^2(d\theta^2 + \sin^2\theta\, d\phi^2), \quad (6.27)$$

where r is to be interpreted as a function of u and v given implicitly by the first equation in (6.26).

Another aspect of the question of singularities can be illustrated with the Schwarzschild metric, as follows (Raychaudhuri 1979, p. 146). Transform the coordinate r in (6.25) to a new coordinate r' given by

$$r - 2m = r'^2. \quad (6.28)$$

This changes (6.25) to the following form:

$$ds = c^2 r'^2/(r'^2 + 2m)\, dt^2 - (r'^2 + 2m)(d\theta^2 + \sin^2\theta\, d\phi^2) - 4(r'^2 + 2m)\, dr'^2. \quad (6.29)$$

Clearly this metric is regular for all values of r' in $0 \leqslant r' \leqslant \infty$. But this is only a part of the space represented by (6.25) with $0 \leqslant r \leqslant \infty$. In (6.29) there would be no singularities of the curvature scalars such as $R_{\alpha\beta\gamma\delta}R^{\alpha\beta\gamma\delta}$ for any values of r'. It is thus not always satisfactory simply to see if the metric components are regular. One way to demand regularity which is physically meaningful is to require that all time-like and null geodesics should be complete in the sense that they can be extended to arbitrary values of their affine parameters. Since time-like and null geodesics give respectively the paths of freely falling that is, in motion under purely gravitational forces) massive and massless particles, this requirement means that the space-time must contain complete histories of such freely falling particles, and that these geodesics should not suddenly come to an end at any point. In fact even this may not be satisfactory as the definition of a regular space-time, as Geroch (1967) has provided an example of a space-time that is geodesically complete (that is, the geodesics can be extended arbitrarily) but one that has a non-geodetic time-like curve (for example an observer propelled by a space-ship, that is, non-gravitational forces) with bounded acceleration which has a finite length. To get over these kinds of difficulties a modified definition of completeness, called b-completeness, has been given by Schmidt (1973).

6.5 An example of a singularity theorem

As indicated earlier there are various forms of singularity theorems, mostly due to Penrose, Hawking and Geroch (see Hawking and Ellis (1973)), which involve elaborate conditions, some of which are quite technical. Roughly speaking, these theorems show that quite reasonable assumptions lead to at least one consequence which is physically unacceptable. We will give here the statement of one of these theorems, due to Hawking and Penrose (1970), which is as follows:

Space-time is not time-like and null geodesically complete if:

(a) $R_{\mu\nu}K^{\mu}K^{\nu} \geqslant 0$ for every non-space-like vector K^{μ}. If the Einstein equations (2.22) are valid, and if K^{μ} is taken to be a unit time-like vector, this condition is readily seen to imply $T_{\mu\nu}K^{\mu}K^{\nu} \geqslant \frac{1}{2}T$. If further $T_{\mu\nu}$ is that for a perfect fluid given by (2.23) and K^{μ} is taken to be the four-velocity u^{μ}, then this condition implies $\varepsilon + 3p \geqslant 0$. For this reason this is sometimes referred to as the energy condition. Physically it is very reasonable.

(b) Every non-space-like geodesic contains a point at which

$$K_{[\mu}R_{\nu]\lambda\sigma[\rho}K_{\tau]}K^{\lambda}K^{\sigma} \neq 0, \qquad [\ ] \text{ implies antisymmetrization,}$$

where K_{μ} is the tangent vector to the geodesics. This is one of the rather technical conditions and it appears that this is true for any general solution of Einstein's equations.

(c) There are no closed time-like curves. Physically this means that no observer can go to his past.

(d) There exists a point p such that the future or past null geodesics from p are focussed by the matter or curvature and start to reconverge. Penrose and Hawking show that observations on the microwave background radiation indicate that this condition is satisfied.

There are actually two alternatives to the condition (d) which are more technical. We refer the interested reader to Hawking and Ellis (1973, p. 266) for an account of this. We thus see that assumptions which are quite reasonable lead to consequences which are physically very strange, such as a particle's world-line suddenly coming to an end, or an observer meeting his past.

6.6 An anisotropic model

To see an example of singularities which is different from the simple Friedmann cases and yet not too complicated, we will consider in this section a model that is homogeneous but anisotropic. It is, in fact, the metric of (2.48) with $A = 1$, and we use X^2, Y^2, Z^2 instead of B, C, D in that equation, so that our metric is as follows:

$$ds^2 = c^2 \, dt^2 - X^2(t) \, dx^2 - Y^2(t) \, dy^2 - Z^2(t) \, dz^2. \qquad (6.30)$$

This metric belongs to Bianchi type I mentioned in Section 6.2. Such models have been studied by Raychaudhuri (1958), Schücking and Heckmann (1958) and others. The case $X = Y$ with dust was considered by Thorne (1967). An account of this model is given in Hawking and Ellis (1973, p. 142).

The fact that the metric (6.30) is homogeneous has been shown at the end of Section 2.2. It is anisotropic because not all directions from a point are equivalent. There are several reasons for studying anisotropic universes. We have mentioned earlier that the universe displays a high degree of isotropy in the present epoch. However, in earlier epochs, perhaps very early ones, there may have been a significant amount of anisotropy. Also, in a realistic situation the singularity in the universe is unlikely to possess the high degree of symmetry that the Friedmann models have. The observed isotropy of the universe needs to be explained and, in the process of seeking this explanation, one must consider more general models of the universe than the Friedmann ones.

We will consider solutions of Einstein's equations for the metric (6.30) for a perfect fluid with zero pressure, that is, dust. We set $G = 1$ and $c = 1$ for this section and the next, and define a function $S(t)$ by $S^3 = XYZ$. A solution of Einstein's equation is given as follows (M, a, b are constants):

$$\left.\begin{array}{l} \varepsilon = 3M/(4\pi S^3), \quad X = S(t^{2/3}/S)^{2\sin a}, \quad Y = S(t^{2/3}/S)^{2\sin(a+\frac{2}{3}\pi)}, \\[2mm] Z = S(t^{2/3}/S)^{2\sin(a+\frac{4}{3}\pi)}, \quad S^3 = \tfrac{9}{2}Mt(t+b). \end{array}\right\} \qquad (6.31)$$

The constant b determines the amount of anisotropy, the value $b = 0$ giving the isotropic Einstein–de Sitter universe (see (3.24)). The constant 'a' determines the direction of most rapid expansion, the domain of 'a' being $-\pi/6 < a < \pi/2$. We have

$$\dot{S}/S = (2/3t)(t + \tfrac{1}{2}b)/(t + b), \quad \dot{X}/X = (2/3t)[t + \tfrac{1}{2}b(1 + 2\sin a)]/(t + b),$$
$$(6.32)$$

the expressions for $\dot{Y}/Y$ and $\dot{Z}/Z$ being obtained by replacing a in $\dot{X}/X$ by $a + \tfrac{2}{3}\pi$ and $a + \tfrac{4}{3}\pi$ respectively. This universe has a highly anisotropic singular state at $t = 0$. For large t it tends to isotropy, in fact to the Einstein–de Sitter universe.

Suppose we follow the time t backwards to the initial singularity. At first there is isotropic contraction. Let $a \neq \frac{1}{2}\pi$. Then $1 + 2 \sin(a + \frac{4}{3}\pi)$ is negative. Thus the collapse in the z-direction halts and is replaced by expansion, the rate of which becomes infinite as t tends to zero. The collapse is monotonic in the x- and y-directions. Consider now the situation forwards from $t = 0$. The matter collapses from infinity in the z-direction, then halts and expands. In the x- and y-directions it expands monotonically. Thus we have here a cigar-shaped singularity. If one could observe the matter far back in time, one would see a maximum red-shift in the z-direction, then the red-shift would decrease to zero (corresponding to the halt), then one would get indefinitely large blue-shifts, the latter occurring in light given off by the matter near $t = 0$.

The case $a = \frac{1}{2}\pi$ is somewhat different. Here we have

$$\dot{X}/X = (2/3t)(t + \tfrac{3}{2}b)/(t + b), \quad \dot{Y}/Y = \dot{Z}/Z = (2/3)(t + b)^{-1}. \tag{6.33}$$

Following time backwards again, the initially isotropic contraction slows down to zero in the y- and z-directions but the collapse is monotonic in the x-direction. Going forwards in time, the rate of expansion of the universe in the y- and z-directions starts from zero but the expansion rate in the x-direction is infinite. This is thus a 'pancake' singularity. There are limiting red-shifts in the y- and z-directions, but no limit to the red-shifts in the x-direction.

6.7 The oscillatory approach to singularities

In this section we consider an interesting approach to singularities developed by Lifshitz and Khalatnikov (1963) and by Belinskii, Khalatnikov and Lifshitz (1970). We study one of the homogeneous spaces that were introduced in Section 6.2, namely, Bianchi type IX, whose structure constants are as follows (see (6.11)):

$$C^1_{23} = C^2_{31} = C^3_{12} = 1. \tag{6.34}$$

Denoting (x^1, x^2, x^3) by (θ, ϕ, ψ), the three vectors $e^{(a)}_m$ (see (6.3) and (6.4)) can be taken as follows:

$$e^{(1)}_m = (\sin \psi, -\cos \psi \sin \theta, 0), \quad e^{(2)}_m = (\cos \psi, \sin \psi \sin \theta, 0), \quad e^{(3)}_m = (0, \cos \theta, 1). \tag{6.35}$$

The metric (6.4) is given as follows, where we have taken $\eta_{ab}(t)$ to be diagonal and set $\eta_{11} = a^2$, $\eta_{22} = b^2$ and $\eta_{33} = c^2$.

$$ds^2 = dt^2 - a^2(\sin \psi \, d\theta - \cos \psi \sin \theta \, d\phi)^2$$
$$- b^2(\cos \psi \, d\theta + \sin \psi \sin \theta \, d\phi)^2 - c^2(\cos \theta \, d\phi + d\psi)^2. \tag{6.36}$$

In the isotropic models studied in Chapter 3, near the singularity the spatial curvature term behaves as R^{-2} whereas the mass-energy density behaves as R^{-3} (for zero pressure) and as R^{-4} (for radiation). (See (3.2a)–(3.2c), (3.15) and (3.40).) Thus in the Friedmann models the curvature terms go to infinity slower than the terms arising from $T_{\mu\nu}$ and the derivatives with respect to time of the metric (that is, $\dot{R}$ terms). This kind of singularity is referred to as a velocity-dominated singularity (Eardly, Liang and Sachs, 1972). In the anisotropic models which are our concern in this section the behaviour near the singularity is dominated by curvature terms as observed by Belinskii and his coworkers and by Misner (1969) and is called the mixmaster singularity.

Thus if we are interested in the behaviour near the initial singularity for the anisotropic metric (6.36), it is sufficient to consider the empty space or vacuum Einstein equations where $T_{\mu\nu} = 0$, for the terms arising from $T_{\mu\nu}$ are negligible in comparison to the other terms. The empty space Einstein equations can be written as follows:

$$(ab\dot{c})^{\cdot}/(abc) = (2a^2b^2c^2)^{-1}[(a^2 - b^2)^2 - c^4], \tag{6.37a}$$

$$(\dot{a}bc)^{\cdot}/(abc) = (2a^2b^2c^2)^{-1}[(b^2 - c^2)^2 - a^4], \tag{6.37b}$$

$$(a\dot{b}c)^{\cdot}/(abc) = (2a^2b^2c^2)^{-1}[(c^2 - a^2)^2 - b^4], \tag{6.37c}$$

$$\ddot{a}/a + \ddot{b}/b + \ddot{c}/c = 0. \tag{6.37d}$$

Here a dot represents differentiation with respect to t. If the right hand sides in (6.37a)–(6.37c) were absent, we would get the following well-known Kasner (1921) solution (of Bianchi type I):

$$a = t^q, \quad b = t^r, \quad c = t^p, \tag{6.38}$$

where p, q, r are constants satisfying

$$p + q + r = p^2 + q^2 + r^2 = 1. \tag{6.39}$$

Suppose now that even when the terms on the right hand sides of (6.37a)–(6.37c) are present, there exist certain ranges of values of t for which the metric is given approximately by (6.38):

$$a \sim t^q, \quad b \sim t^r, \quad c \sim t^p. \tag{6.40}$$

Then from (6.37d) we get

$$p^2 + q^2 + r^2 = p + q + r. \tag{6.41}$$

It is readily verified that not all the three expressions on the right hand sides of (6.37a)–(6.37c) can be positive, that is, one of these at least must be negative. From this it follows, substituting (6.40) into the left hand sides of (6.37a)–(6.37c), that at least one of the expressions $p(p + q + r - 1)$, $q(p + q + r - 1)$, $r(p + q + r - 1)$ must be negative. The possibility that p, q, r are all positive with $p + q + r - 1$ negative is inadmissible because it contradicts (6.41) (for in this case we must have $0 < p < 1$, $0 < q < 1$, $0 < r < 1$, so that $p^2 < p$, $q^2 < q$, $r^2 < r$, and (6.41) becomes impossible). Thus at least one of the indices p, q, r is negative. This implies that the length along at least one direction shrinks while (since $p + q + r > 0$ from (6.41)) the spatial volume, which is determined by the product $(abc)^2$, expands. In fact (6.37a)–(6.37c) do not allow two of the exponents p, q, r to be negative at the same time.

We suppose that p is negative and $q < r$. Then (6.40) implies that for small t, a and b can be neglected in comparison with c. We now define new dependent variables α, β, γ and a new independent variable τ by the following relations:

$$a = \exp(\alpha), \quad b = \exp(\beta), \quad c = \exp(\gamma); \quad dt/d\tau = abc. \tag{6.42}$$

These transformations, together with the approximations introduced above, enable us to write (6.37a)–(6.37c) as follows:

$$\gamma'' = -\tfrac{1}{2}\exp(4\gamma), \tag{6.43a}$$

$$\alpha'' = \beta'' = \tfrac{1}{2}\exp(4\gamma), \tag{6.43b}$$

where a prime denotes differentiation with respect to τ. Equation (6.43a) is in the form of the equation of motion of a particle which is moving in a potential well which is exponential. The 'velocity' γ' thus changes sign corresponding to a change from a region where c is decreasing to one where c is increasing. Belinskii *et al.* assume that the right hand sides of (6.37a)–(6.37c) are small enough at a certain epoch such that $p + q + r$ is nearly unity and one has the Kasner solution with

$$abc = wt, \quad \tau = w^{-1}\log t + \text{constant}, \tag{6.44}$$

where w is a constant. Equations (6.43a) and (6.43b) can then be integrated as follows:

$$a^2 = a_0^2[1 + \exp(4pw\tau)]\exp(2qw\tau), \tag{6.45a}$$

$$b^2 = b_0^2[1 + \exp(4pw\tau)]\exp(2rw\tau), \tag{6.45b}$$

$$c^2 = 2|p|[\cosh(2wp\tau)]^{-1}, \tag{6.45c}$$

where we have chosen the integration constants so that as τ tends to infinity, a, b, c go to the assumed Kasner solution with a negative p. We get the following asymptotic values of a, b, c as τ tends to infinity and minus infinity respectively:

As $\tau \to \infty$, $a \sim \exp(qw\tau)$, $b \sim \exp(rw\tau)$, $c \sim \exp(pw\tau)$,   (6.46a)

As $\tau \to -\infty$, $a \sim \exp[w(q + 2p)\tau]$, $b \sim \exp[w(r + 2p)\tau]$,

$c \sim \exp(-pw\tau)$.   (6.46b)

In (6.46a) we have $w\tau \sim \log t$ while in (6.46b), $w(1 + 2p)\tau \sim t$. In the second of these limits, that is in (6.46b), transforming back to t from τ (with $w(1 + 2p)\tau = t$), we get

$$a \sim t^{q'}, \quad b \sim t^{r'}, \quad c \sim t^{p'}, \quad (6.47)$$

where

$$p' = -p/(1 + 2p) > 0, \quad (6.48a)$$

$$q' = (2p + q)/(1 + 2p) < 0, \quad (6.48b)$$

$$r' = (r + 2p)/(1 + 2p) > 0. \quad (6.48c)$$

This behaviour is different from that existing in the limit $\tau \to \infty$ which is given by (6.40), in the sense that the exponent in c has changed from negative to positive, while that of a has become negative (that is, q is positive but q' negative). Thus the a- and c-axes have interchanged their expanding and contracting behaviours. This indicates that, as we move towards the singularity, distances along two of the axes oscillate while that along the third axis decreases monotonically. This happens in successive periods which are called 'eras'. On going from one era to the next, the axis along which distances decrease monotonically changes to another one. Asymptotically the order in which this change occurs becomes a random process (Landau and Lifshitz 1975). One has a particularly long era if (p, q, r) corresponds to the triplet $(1, 0, 0)$. In this case there are no particle horizons (see Section 3.7) in the direction for which the index is unity, since $\int_0 t^{-1}\, dt$ diverges. In the course of evolution this particular direction also changes and this phenomenon may lead to effective abolition of all particle horizons. This was one of the motivations of the mixmaster model of Misner which was thought to provide the solution to the 'horizon' problem mentioned in Chapter 1, that is, to explain why the universe is so isotropic and homogeneous. But this model did not provide a solution to the problem, although some interesting insights were gained. This completes our brief exposition of singularities in cosmology. For more details of the material presented in this chapter, we refer to the books by Hawking and Ellis (1973), Raychaudhuri (1979),

Landau and Lifshitz (1979) and the papers cited in this chapter. There have also been interesting inhomogeneous exact cosmological solutions following the work of Szekeres (1975); see, for example, the papers by Szafron (1977), Szafron and Wainwright (1977), Wainwright (1979), Wainwright and Marshman (1979), Wainwright, Ince and Marshman (1979), Wainwright and Goode (1980), Wainwright (1981), and Goode and Wainwright (1982). It is, however, beyond the scope of this book to consider these models.

7

The early universe

7.1 Introduction

As mentioned in Chapter 1, the 'cosmic background radiation' discovered originally by Penzias and Wilson in 1965 provides evidence that the universe must have gone through a hot dense phase. We have also seen that the Friedmann models (described in Chapter 3), if they are regarded as physically valid, predict that the density of mass-energy must have been very high in the early epochs of the universe. In fact, of course, the Friedmann models imply that the mass-energy density goes to infinity as the time t approaches the 'initial moment' or 'the initial singularity', at $t = 0$. This is what is referred to as the 'big bang', meaning an explosion at every point of the universe in which matter was thrown asunder violently, from an infinite or near infinite density. However, the precise nature of the physical situation at $t = 0$, or the situation *before* $t = 0$ (or whether it is physically meaningful to talk about any time *before* $t = 0$) – these sorts of questions are entirely unclear. In this and the following chapter we shall try to deal partly with some questions of this kind. In the present chapter we simply *assume* that there was a catastrophic event at $t = 0$, and try to describe the state of the universe from about $t = 0.01$ s until about $t =$ one million years. This will be our definition of the 'early universe', which specifically excludes the first hundredth of a second or so, during which, as we shall see in the next chapter, and as speculations go, events occurred which are of a very different nature from those occurring in the 'early universe' according to the definition given here.

In this section we shall describe qualitatively the state of the early universe and in the following sections we shall provide a more quantitative account of this state. The description given in this section is derived largely from that given in Weinberg's book (1977, 1983). As indicated in Fig. 1.3, the spectrum of the cosmic background radiation peaks at slightly under

0.1 cm. Penzias and Wilson made their original observation at 7.35 cm. Since that time there have been many observations, both ground-based and above the atmosphere, which confirm the black-body nature of the radiation, with a temperature of about 2.7 K. Below about 0.3 cm, the atmosphere becomes increasingly opaque, so such observations have to be carried out above the atmosphere. Although at times there have been slight doubts, it is now generally agreed that the cosmic background radiation is indeed the remnant of the radiation from the early universe, which has been red-shifted, that is, reduced in temperature to 2.7 K. As we shall see later in more detail, the temperature of the cosmic background radiation provides us with an important datum about the universe, that there are about 1000 million photons in the universe for every nuclear particle; by the latter we mean protons and neutrons, or 'baryons'. There is some uncertainty in this figure, but we shall use this figure for the time being, and later explain the possible modification.

To describe the state of the early universe we choose several instants of time, which are referred to by Weinberg as 'frames', as if a movie had been made and we were looking at particular frames in this movie. These instants of time are chosen so that major changes take place near those times. In the following we describe the physical state of the universe at these instants, or frames. (The values of the temperature, time etc. are slightly different from those in Weinberg (1977. 1983) to conform with subsequent calculations in this book.)

(i) First frame

This is at $t = 0.01$ s, when the temperature is around 10^{11} K, which is well above the threshold for electron–positron pair production. The main constituents of the universe are photons, neutrinos and antineutrinos, and electron–positron pairs. There is also a small 'contamination' of neutrons, protons and electrons. The energy density of the electron–positron pairs is roughly equal to that of the neutrinos and antineutrinos, both being $\frac{7}{4}$ times the energy density of the photons. The total energy density is about 21×10^{44} eV 1^{-1}, or about 3.8×10^{11} g cm^{-3}. The characteristic expansion time of the universe (that is, the reciprocal of Hubble's 'constant' at that instant, which is the age of the universe if the rate of expansion had been the same from the beginning as at that instant) is 0.02 s. The neutrons and protons cannot form into nuclei, as the latter are unstable. The spatial volume of the universe would be either infinite or, if it is one of the finite models, say with density twice the critical density, its circumference would be about 4 light years.

(ii)   Second frame

This is at $t = 0.12$ s, when the temperature has dropped to about 3×10^{10} K. No qualitative changes have occurred since the first frame. As in the first frame, the temperature is above electron–positron pair threshold, so that these particles are relativistic, and the whole mixture behaves more like radiation than matter, with the equation of state given nearly by $p = \frac{1}{3}\varepsilon$. The total density is about 3×10^7 g cm^{-3}. The characteristic expansion time is about 0.2 s. No nuclei can be formed yet, but the previous balance between the numbers of neutrons and protons, which were being transformed into each other through the reaction $n + v \rightleftarrows p + e^-$, is beginning to be disturbed as neutrons now turn more easily into the lighter protons than vice versa. Thus the neutron–proton ratio becomes approximately 38% neutrons and 62% protons. The thermal contact (see below) between neutrinos and other forms of matter is beginning to cease.

(iii)   Third frame

This is at $t = 1.1$ s, when the temperature has fallen to about 10^{10} K. The thermal contact between the neutrinos and other particles of matter and radiation ceases. Thermal contact is here taken to mean the conversion of electron–positron pairs into neutrino–antineutrino pairs and vice versa, the conversion of neutrino–antineutrino pairs into photons and vice versa, etc. Henceforth neutrinos and antineutrinos will not play an active role, but only provide a contribution to the overall mass-energy density. The density is of the order of 10^5 g cm^{-3} and the characteristic expansion time is a few seconds. The temperature is near the threshold temperature for electron–positron pair production, so that these pairs are beginning to annihilate more often to produce photons than their creation from photons. It is still too hot for nuclei to be formed and the neutron–proton ratio has changed to approximately 24% neutrons and 76% protons.

(iv)   Fourth frame

This is approximately at $t \approx 13$ s, when the temperature has fallen to about 3×10^9 K. This temperature is below the threshold for electron–positron production so most of these pairs have annihilated. The heat produced in this annihilation has temporarily slowed down the rate of cooling of the universe. The neutrinos are about 8% cooler than the photons, so the energy density is a little less than if it were falling simply as the fourth power of the temperature (recall that according to the

Stefan–Boltzmann law $\varepsilon = \sigma T^4$ ergs cm^{-3}, where $\sigma = 7.564\,64 \times 10^{-15}$, and T is the temperature in K). The neutron–proton balance has shifted to about 17% neutrons and 83% protons. The temperature is low enough for helium nuclei to exist, but the lighter nuclei are unstable, so the former cannot be formed yet. By helium nuclei we mean alpha particles, He^4, which have two protons and two neutrons. The expansion rate is still very high, so only the light nuclei form in two-particle reactions, as follows: $p + n \rightarrow D + \gamma$, $D + p \rightarrow He^3 + \gamma$, $D + n \rightarrow H^3 + \gamma$, $He^3 + n \rightarrow He^4 + \gamma$, $H^3 + p \rightarrow He^4 + \gamma$. Here D denotes deuterium, which has one neutron and one proton, He^3 is helium-3, an isotope of helium with two protons and one neutron, H^3 is tritium, an isotope of hydrogen with one proton and two neutrons, and γ stands for one or more photons. Although helium is stable, the lighter nuclei mentioned here are unstable at this temperature, so helium formation is not yet possible, as it is necessary to go through the above intermediate steps to form helium. The energy required to pull apart the neutron and proton in a D nucleus, for example, is one-ninth that required to pull apart a nucleon (neutron or proton) from an He^4 nucleus. In other words, the binding energy of a nucleon in deuterium is one-ninth that in an He^4 nucleus.

(v) Fifth frame

This is about 3 min after the first frame when the temperature is about 10^9 K, which is approximately 70 times as hot as the centre of the Sun. The electron–positron pairs have disappeared, and the contents of the universe are mainly photons and neutrinos plus, as before, a 'contamination' of neutrons, protons and electrons (whose numbers are much smaller than the number of photons, by a ratio of about $1:10^9$), which will eventually turn into the matter of the present universe. The temperature of the photons is about 35% higher than that of the neutrinos. It is cool enough for H^3, He^3 and He^4 nuclei to be stable, but the deuterium 'bottleneck' is still at work so these nuclei cannot be formed yet. The beta decay of the neutron into a proton, electron and anti-neutrino is becoming important, for this reaction has a time scale of about 12 min. This causes the neutron–proton balance to become 14% neutrons and 86% protons.

A little later than the fifth frame the temperature drops enough for deuterium to become stable, so that heavier nuclei are quickly formed, but as soon as He^4 nuclei are formed other bottlenecks operate, as there are no stable nuclei at that temperature with five or eight particles. The exact temperature depends on the number of photons per baryon; if this number is 10^9 as assumed before, then the temperature is about 0.9×10^9 K, and these events take place at some time between $t = 3$ min and $t = 4$ min. Nearly all the neutrons are used up to make He^4, with very few heavier nuclei due to the other bottlenecks mentioned. The neutron–proton ratio is about 12% or 13% neutrons to 88% or 87% protons, and it is frozen at this value as the neutrons have been used up. As the He^4 nuclei have equal numbers of neutrons and protons, the proportion of helium to hydrogen nuclei (the latter being protons) by weight is about 24% or 26% helium and 76% or 74% hydrogen. This process, by which heavier nuclei are formed from hydrogen, is called *nucleosynthesis*. If the number of photons per baryon is lower (that is, if the baryon:photon ratio is higher), then nucleosynthesis begins a little earlier, and slightly more He^4 nuclei are formed than 24% or 26% by weight.

(vi)   Sixth frame

This is approximately at $t \simeq 35$ min, when the temperature is about 3×10^8 K. The electrons and positrons have annihilated completely, except for the small number of electrons left over to neutralize the protons. It is assumed throughout that the charge density in any significant volume of the universe is zero. The temperature of the photons is about 40% higher than the neutrino temperature, and will remain so in the subsequent history of the universe. The energy density is about 10% the density of water, of which 31% or so is contributed by neutrinos and the rest by photons. The density of 'matter' (that is, of the nuclei and protons etc.) is negligible in comparison to that of photons and neutrinos. The characteristic expansion time of the universe is about an hour and a quarter. Nuclear processes have then stopped, the proportion of He^4 nuclei being anywhere between 20% and 30% depending on the baryon: photon ratio (see Fig. 7.1).

We see from the preceding discussion that the proportion of helium nuclei formed in the early universe was anything from 20–30% by weight, with very few heavier nuclei due to the five- and eight-particle bottlenecks. For the nucleosynthesis process to take place one needs temperatures of the order of a million degrees. After the temperature dropped below about a million degrees in the early universe, the only place in the later universe where similar temperatures exist would be the centre of stars. It can be shown that no significant amount of helium (compared to the 20–30% of the early universe) could have been created in the centre of stars. This follows from the fact that such a significant amount of helium formation would have released so much energy into the interstellar and intergalactic space, that it would be inconsistent with the amount of radiation actually given off since the time of star and galaxy formation, an amount of which can be calculated from the average absolute luminosity of stars and galaxies, which are known, and the time scale during which these have existed, which is from soon after the recombination era (see below). Thus if the above picture is reasonable, there should be approximately 20–30% helium nuclei in the present universe, most of the rest being predominantly hydrogen, with a small amount of heavier nuclei. This is indeed found to be the case. We shall have more to say about this later in this chapter.

We have seen that the time, temperature and the extent of nucleosynthesis depends on the density of nuclear particles compared to photons. The amount of deuterium that was produced by nucleosynthesis in the early universe and the amount that survives and should be observable today, depends very sensitively on the nuclear particle to photon ratio. As an illustration of this, we give in Table 7.1 the abundance of deuterium as worked out by Wagoner (1973) for three values of the photon : nuclear particle ratio. We shall have more to say about deuterium later in this chapter.

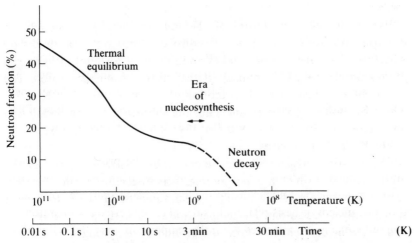

Fig. 7.1. This diagram describes the neutron–proton ratio in the early universe. The period of 'thermal equilibrium' is one in which all particles and radiation are in equilibrium and the neutron–proton ratio depends on the mass difference between these particles. The 'era of nucleosynthesis' is the period when lighter nuclei, predominantly helium, are being formed. The dashed portion indicates that if the neutrons had not been incorporated into nuclei they would have decayed through beta decay (Weinberg, 1977).

Table 7.1 *Abundance of deuterium and the photon:baryon ratio.*

Photons:nuclear particle	Deuterium abundance (parts/10^6)
100 million	0.000 08
1 000 million	16
10 000 million	600

We have seen that after the first few minutes the only particles left in the universe were photons, neutrinos, neutrons, protons and electrons. The latter two particles are charged ones, and in their free state they could scatter photons freely. As a result the 'mean free path' of photons, that is, the average distance that a photon travels in between scatterings by two charged particles, was small compared to the distance a photon would travel during the characteristic expansion time of the universe for that period, if it were unimpeded. This is what is meant by the matter and radiation being in equilibrium, as there is free exchange of energy between the two. Thus the universe, during the period that protons and electrons were free particles, was opaque to electromagnetic radiation.

Eventually the temperature of the universe was cool enough for electrons and protons to form stable hydrogen atoms in their ground state when they combined. Now it takes about 13.6 eV to ionize a hydrogen atom completely, that is, pull apart the electron from the proton. The energy of a particle in random motion at a temperature of T K is kT, where k is Boltzmann's constant. Thus the temperature corresponding to an energy of 13.6 eV is k^{-1} times 13.6, where k^{-1} is approximately 11 605 K eV^{-1}. This gives about 1.576×10^5 K as the temperature at which a hydrogen atom is completely ionized. However, even in the excited states, in which it is not ionized, a hydrogen atom can effectively scatter photons. Thus it is only in the ground state that it ceases to interact significantly with photons. The temperature at which the primeval protons and electrons combined to form the ground state hydrogen atoms was about 3000–4000 K, which occurred a few hundred thousand years after the big bang. This era is referred to as 'recombination' (a singularly inappropriate term, as Weinberg remarks, as the electrons and protons were never in a combined state before!). After this period the universe became transparent to electromagnetic radiation, that is, the mean free path of a photon became much longer than the distance traversed in a characteristic expansion time of the period. This is the reason we get light, which has hardly been impeded, except for the red-shift, from galaxies billions of light years away.

7.2   The very early universe

In the last section we discussed qualitatively the early universe which we defined to begin at about $t = 0.01$ s. In this section we shall give a qualitative and speculative discussion of the very early universe, which we take to be the first hundredth of a second or so. As mentioned in Chapter 1 and also earlier in this chapter, there have been elaborate speculations about the very early universe. We shall discuss these speculations in some detail in the next chapter, where we shall give a quantitative discussion wherever possible. Some of the remarks made in this section may have to be qualified in the next chapter.

As we shall see more clearly in the next chapter, the very early universe involves elementary particles and their interactions in an intimate way, much more so than the early universe. For this reason it is necessary to know something about these particles. Table 7.2 gives the classification and properties of the more common elementary particles. As is well known from quantum field theory, which is the theory describing the interactions of these particles, the interactions can be described in a picturesque way by Feynman diagrams, which give the amplitudes for various processes to certain order in the coupling constant. Three such diagrams are given in Fig. 7.2, corresponding to electromagnetic, strong and weak interactions. These interactions and gravitation are described in Table 7.3. There is a considerable amount of uncertainty in our knowledge of the first hundredth of a second of the universe. This stems partly from our inadequate knowledge of the strong interactions of elementary particles. As we go to higher temperatures than the first frame temperature of about 10^{11} K, nearer $t = 0$, there would be copious production of hadrons, and it becomes difficult to describe the nature of matter at these temperatures for this reason, as the hadrons take part in strong interactions, whose precise nature is not known. There are two views of the nature of matter at such energies. The first one, which is not in favour at present, says that there are no 'elementary' hadrons but that every hadron is in a sense a composite of all other hadrons. In this case, as the temperature increases, the energy available goes into producing more massive hadrons, and not into the random motion of the constituent particles. As there is no limit to the mass of these 'elementary' hadrons, there is a maximum possible temperature, around 2×10^{12} K, even though the density goes to infinity. The idea of this 'nuclear democracy' was mainly due to G. Chew; the

Table 7.2 In this table are listed the more common elementary particles. Particles and their antiparticles have the same mass, same life-time and opposite charges, so they are listed in the same line. A symbol with a bar over it denotes an antiparticle; thus $\bar{\nu}_\mu$ is the muon-antineutrino. Leptons take part in weak interactions but not in strong interactions. All hadrons take part in strong interactions: they are made up of mesons (which are bosons) and baryons (which are fermions). All hadrons also take part in weak interactions. Baryons other than the proton and the neutron are called hyperons.

	Particle	Symbol	Charge (in units of proton charge)	Mass (MeV)	Life-time (s)	Spin (in units of $\hbar$)
	Photon	γ	0	0	infinite	1
Leptons	Neutrino	$\nu_e, \bar{\nu}_e$	0	less than 0.001	infinite (?)	$\tfrac{1}{2}$
		$\nu_\mu, \bar{\nu}_\mu$	0	less than 0.001	infinite (?)	$\tfrac{1}{2}$
	Electron	$e^\pm$	± 1	0.51	infinite	$\tfrac{1}{2}$
	Muon	$\mu^\pm$	± 1	105.66	2.2×10^{-6}	$\tfrac{1}{2}$
Mesons	Pion	$\pi^\pm$	± 1	139.57	2.6×10^{-8}	0
		π^0	0	134.97	0.84×10^{-16}	0
	Kaon	$\kappa^\pm$	± 1	493.71	1.24×10^{-8}	0
		$\kappa^0, \bar{\kappa}^0$	0	497.71	0.88×10^{-10}	0
	Eta	η	0	548.8	2.50×10^{-17}	0
Hadrons	Proton	$p, \bar{p}$	± 1	938.26	infinite (?)	$\tfrac{1}{2}$
	Neutron	$n, \bar{n}$	0	939.55	918	$\tfrac{1}{2}$
Baryons	Lambda hyperon	$\Lambda, \bar{\Lambda}$	0	1115.59	2.52×10^{-10}	$\tfrac{1}{2}$
	Sigma hyperon	$\Sigma^+, \bar{\Sigma}^+$	± 1	1189.42	8.00×10^{-11}	$\tfrac{1}{2}$
	Sigma hyperon	$\Sigma^0, \bar{\Sigma}^0$	0	1192.48	less than 10^{-14}	$\tfrac{1}{2}$
	Sigma hyperon	$\Sigma^-, \bar{\Sigma}^-$	± 1	1197.34	1.48×10^{-10}	$\tfrac{1}{2}$
	Cascade hyperon	$\Xi^0, \bar{\Xi}^0$	0	1314.7	2.98×10^{-10}	$\tfrac{1}{2}$
	Cascade hyperon	$\Xi^-, \bar{\Xi}^-$	± 1	1321.3	1.67×10^{-10}	$\tfrac{1}{2}$
	Omega hyperon	$\Omega^-, \bar{\Omega}^-$	± 1	1672	1.3×10^{-10}	$\tfrac{1}{2}$

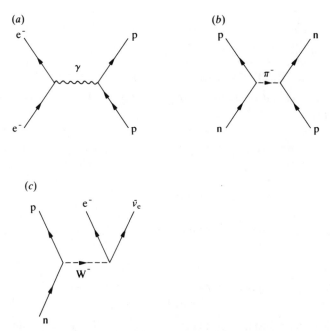

Fig. 7.2. This figure illustrates how forces are mediated by the exchange of particles. In (*a*) an electron (e$^-$) and a proton (p) interact by exchanging a photon (γ). In (*b*) a neutron becomes a proton by emitting a π-meson, which is then absorbed by another proton which subsequently becomes a neutron. In (*c*) the beta decay of a neutron is caused by the emission of an intermediate vector meson W$^-$ which decays into an electron and an electron-antineutrino.

maximum temperature in hadron physics was pointed out by R. Hagedorn (see, for example, Huang and Weinberg (1970)).

In the second view of particle physics all hadrons are made of a few fundamental constituents, known as quarks. They come in six varieties, known as 'flavours', these being the up, down, strange, charmed, top and bottom quarks, represented respectively by the letters u, d, s, c, t, b (the latter two are sometime called 'truth' and 'beauty'). There are also the corresponding antiquarks denoted by $\bar{u}$, $\bar{d}$, etc. These quarks have fractional charges (see Table 7.4) and each flavour comes in three states called 'colours', usually referred to as yellow, blue and red, with the corresponding antiquarks being antiyellow, etc. Within a baryon or a meson the quarks interact with each other by exchanging still other fundamental particles called 'gluons' of which there are eight kinds, depending on their colour composition. The hadrons are 'colourless',

Table 7.3 *This table gives some properties of the four kinds of forces encountered in nature so far, namely gravitational, electromagnetic, strong (nuclear) and weak forces. 'Particles exchanged' means the particles through the exchange of which the corresponding force is mediated. The 'graviton' is a hypothetical particle through the exchange of which gravitational forces are mediated.*

	Gravitational force	Electromagnetic force	Strong (nuclear) force	Weak force
Range	Infinite	Infinite	$10^{-13}-$ 10^{-14} cm	Less than 10^{-14} cm
Examples	Astronomical forces	Atomic forces	Nuclear forces	Beta decay of neutron
Strength	10^{-39}	$\frac{1}{137}$	1	10^{-5}
Particles acted upon	Everything	Charged particles	Hadrons	Hadrons and leptons
Particles exchanged	Gravitons (?)	Photons	Hadrons	Intermediate vector bosons

Table 7.4 *In one form of the grand unified theories there is a correspondence between leptons and quarks as shown in this table. See the text for the meaning of the quark symbols. The τ^- refers to the τ-lepton and v_τ is the corresponding neutrino. Each of the quarks come in three 'colours'.*

	Leptons		Quarks	
	Symbol	Charge	Symbol	Charge
First generation	v_e	0	u	$+\frac{2}{3}$
	e^-	-1	d	$-\frac{1}{3}$
Second generation	v_μ	0	c	$+\frac{2}{3}$
	μ^-	-1	s	$-\frac{1}{3}$
Third generation	v_τ	0	t	$+\frac{2}{3}$
	τ	-1	b	$-\frac{1}{3}$

being composite of quarks of all three colours, or quarks of a certain colour and its anticolour.

The Glashow–Weinberg–Salam theory gives a unified description of the weak and electromagnetic interactions, according to which above a certain energy both interactions are similar and have the same strength. There have been attempts at unifying these with the strong interactions – the Grand Unified Theories – but these have not been so successful.

Although there are strong indications that hadrons are made of quarks, no quarks have been observed yet. A satisfactory explanation of this phenomenon has not been found, although there are some hints in the property of 'asymptotic freedom', which is a consequence of the gauge theory which is thought to describe the interactions of quarks and gluons. This property indicates that the strength of the interaction between two quarks becomes negligible when they are close together, and correspondingly the strength increases when they are far apart. Thus if one attempts to detach a quark from other quarks in a baryon, say, the energy required eventually becomes so great that a quark–antiquark pair is formed, so that these combine with the existing quarks to form two hadrons, and one does not get a free quark. Thus in the quark model, in the very early universe the quarks must have been very close to each other and so behaved essentially as free particles. As the universe cooled, every quark must have either annihilated with another quark to produce a meson, or else formed a part of a neutron or a proton. In this case both the temperature and the density tends to infinity as t tends to zero.

There is a possibility that the universe may have suffered a phase transition as the universe cooled, somewhat like the freezing of water. At this phase transition, the electromagnetic and weak interactions may have become different. In the Glashow–Weinberg–Salam unification of electromagnetism and the weak interactions, the basic theory used is a gauge theory. One way of looking at this unification is as follows. Electromagnetic interactions between charged particles are mediated via the photon, which is a massless spin 1 particle (see Fig. 7.2(a)). The weak interactions are mediated by massive intermediate vector bosons, the $W^{\pm}$ and Z^0 particles, which are spin 1 particles with masses of about 80 and 90 proton masses respectively. Now at energies which are much higher than the energies represented by these masses, the intermediate boson masses can be neglected so that the weak interactions can be considered as being mediated by massless spin 1 particles. This is akin to the electromagnetic interactions so that at these energies the two interactions behave in a similar manner. It was shown in 1972 by Kirzhnits and Linde that, in fact, gauge theories exhibit a phase transition at a critical temperature of about 3×10^{15} K. Above this temperature the unity between the electromagnetic and weak interactions that is incorporated in the Glashow–Weinberg–Salam model was manifest. Below this temperature the weak interactions became short range while the electromagnetic interactions continued to be long range (these are characteristics of interactions which are mediated respectively by massive and massless particles). When water freezes, a certain symmetry is lost, for example, ice crystals at any point do not possess the same rotational symmetry as liquid water. Secondly, the frozen ice is separated into different domains with different crystal structures. It is conceivable that after the phase transition at some critical temperature the universe has different domains in which the erstwhile symmetry between the electromagnetic and weak interactions is broken in different manners, and that we live in one of these domains. There may remain in the universe zero-, one- or two-dimensional 'defects' from the time of the phase transition.

There is also the possibility that at higher temperatures there may have been symmetry between all three of the microscopic interactions – the weak, electromagnetic and strong interactions, and at yet higher temperatures the weakest of the forces, gravitation, may also have been included in this symmetry. At superhigh temperatures the energies of particles in thermal equilibrium may be so large that the gravitational force between them may be comparable to any other force. This may occur at 10^{32} K, at about 10^{-43} s after $t = 0$. In this situation the horizon would be at a distance less than what we regard as the radius of the particles, that is, crudely speaking, each particle would be as big as the observable universe!

Just as neutrinos and then photons decoupled from matter and continued to form a part of the 'background' radiation, so at a much earlier time gravitational radiation would have also decoupled and there must also be present cosmic background gravitational radiation with a temperature of about 1 K. If it could be detected, it would give us information about a much earlier epoch of the universe than the photon or the neutrino background radiation. However, this is far beyond present technology, as gravitational radiation has not yet been detected in any form.

After the above qualitative descriptions of the early and the very early universe, we go on to more quantitative descriptions in this and the next chapter. The rest of this chapter is based mainly on Weinberg (1972, 1983), Schramm and Wagoner (1974), Bose (1980), and Gautier and Owen (1983).

7.3   Equations in the early universe

We see from (3.15) and (3.40) that in the matter-dominated and radiation-dominated situations the mass-energy density varies as R^{-3} and R^{-4} respectively. Thus in these situations εR^2 varies as R^{-1} and R^{-2} respectively. We know that in all the Friedmann models R starts from the value zero at $t = 0$. Thus in any case εR^2 tends to infinity as t tends to zero. This shows (see (2.109a) and (3.2a)–(3.2c)) that near $t = 0$, that is, in the early universe, one can approximate the evolution of R for all three values of k by the same equation, (3.2b), that is, as follows:

$$\dot{R}^2 = (8\pi G/3)\varepsilon R^2/c^2. \tag{7.1}$$

This in turn shows that the initial behaviour of R is independent of whether the universe is open or closed. We have seen that the early universe is dominated either by radiation or radiation and highly relativistic particles. For these the equation of state is $p = \frac{1}{3}\varepsilon$, so that we get the mass-energy density ε behaving as R^{-4}. Now according to the Stefan–Boltzmann law the energy-density of radiation varies as T^4, where T is the absolute temperature. Thus the temperature of the radiation (and relativistic matter) in the early universe varies as R^{-1}. After the decoupling of matter and radiation the temperature of the radiation continues to decrease as R^{-1}. For a short period there is modification of this behaviour (see below).

Equation (7.1) has the consequence that in the early universe R behaves as $t^{1/2}$, since the equation of state is $p = \frac{1}{3}\varepsilon$ (see (3.47)). If the early universe had been matter-dominated, R would have varied as $t^{2/3}$ (see (3.45)).

The early universe, which is radiation-dominated, can thus be characterized by connecting values of R, ε, T at any two instants of time t_1 and t_2, as follows:

$$R_1/R_2 = t_1^{1/2}/t_2^{1/2} = \varepsilon_2^{1/4}/\varepsilon_1^{1/4} = T_2/T_1. \tag{7.2}$$

provided no major changes take place in the constitution of the contents, such as electron–positron annihilation. For example, for the whole of the radiation-dominated period after the electron–positron annihilation, the energy density is given as follows:

$$\varepsilon = 1.22 \times 10^{-35} T^4 \text{ g cm}^{-3}, \tag{7.3}$$

(see (7.23) below) where here as elsewhere, T denotes absolute temperature.

7.4   Black-body radiation and the temperature of the early universe

Although the properties of black-body radiation are well known, we give here a brief summary for completeness. The energy density of black-body radiation in a range of wavelengths from λ to $\lambda + d\lambda$ is given by the Planck formula as follows:

$$du = (8\pi hc/\lambda^5)\, d\lambda[\exp(hc/kT\lambda) - 1]^{-1}, \tag{7.4}$$

where k is Boltzmann's constant (1.38×10^{-16} erg K^{-1}), h is Planck's constant (6.625×10^{-27} erg s). For long wavelengths, neglecting higher powers of λ^{-1}, (7.4) reduces to

$$du = (8\pi kT/\lambda^4)\, d\lambda, \tag{7.5}$$

which is the Rayleigh–Jeans formula. If this formula is continued to $\lambda = 0$, one gets an infinite energy density. The maximum of du in the Planck formula (7.4) occurs at the value of λ given by the following equation:

$$5kT\lambda[\exp(hc/kT\lambda) - 1] = hc \exp(hc/kT\lambda). \tag{7.6}$$

The solution of this transcendental equation is given approximately as follows:

$$\lambda_0 = 0.201\,405\,2hc/kT, \tag{7.7}$$

which shows that the wavelength at which the maximum occurs is inversely proportional to the temperature. The total energy at temperature T is obtained by integrating (7.4) over all wavelengths:

$$\begin{aligned}
u &= \int_0^\infty (8\pi hc/\lambda^5)[\exp(hc/kT\lambda) - 1]^{-1}\, d\lambda \\
&= \int_0^\infty (8\pi h\nu^3/c^3)[\exp(h\nu/kT) - 1]^{-1}\, d\nu,
\end{aligned} \tag{7.8}$$

where in the last step we have expressed the integral in terms of the frequency $\nu = c/\lambda$. The result of the integration is as follows:

$$u = 8\pi^5(kT)^4/15h^3c^3 = 7.5641 \times 10^{-15} T^4 \text{ erg cm}^{-3}. \tag{7.9}$$

Since a photon has energy $h\nu = hc/\lambda$, the number density of photons is given as follows, for wave-lengths from λ to $\lambda + d\lambda$:

$$dN = du/h\nu = \lambda\, du/hc = (8\pi/\lambda^4)[\exp(hc/kT\lambda) - 1]^{-1}, \tag{7.10}$$

and the number density of photons is

$$N = \int_0^\infty dN = 60.421\,98(kT)^3/(hc)^3 = 20.28\,T^3 \text{ photons cm}^{-3} \quad (7.11)$$

and the energy per photon is

$$u/N = 3.73 \times 10^{-16}T. \qquad (7.12)$$

Equation (7.11) enables us to make a rough estimate of the photon: baryon ratio mentioned earlier. In the present universe, almost all the photons are in the cosmic background radiation – the number of photons that make up the radiation from stars and galaxies is negligible in comparison. Assuming the background radiation to have temperature 2.7 K, (7.11) then gives about 400 photons cm^{-3} as the present number density. We have seen earlier that there is an uncertainty in the present matter density of the universe. Assuming H_0 to be 50 km s^{-1} Mpc^{-1}, (3.9) gives 4.9×10^{-30} g cm^{-3} as the critical density. Let us suppose that the actual density is anywhere from 0.1 to 2 times the critical density. Since the matter is predominantly in baryons, this makes the baryon number density lie approximately between 0.3×10^{-6} and 6×10^{-6} per cubic centimetre (using the fact that a proton has mass 1.67×10^{-24} g). This implies that the ratio of baryons to photons lies approximately between 0.75×10^{-9} and 1.5×10^{-8}. Taking reciprocals, the ratio of photons to baryons is between 1.33×10^9 and 6.6×10^7. Although there is some uncertainty, the cosmic background radiation thus provides us with the useful piece of information of the approximate ratio of the numbers of photons and baryons. This number does not change as the universe evolves, unless it has gone through a stage which produces significant numbers of photons through friction and viscosity, which seems unlikely if the standard model is correct. A knowledge of the photon:baryon number ratio enables us to infer the rate at which nucleosynthesis proceeded in the early universe, and to compare these predictions with the existing abundances of the nuclei. Although there are uncertainties in various stages, the above considerations do provide information about the different pieces in the jigsaw.

There is another way to look at the increase of wavelength of the background photons as the universe expands. Let R change by a factor f. Then the wavelength of a typical ray of light will also change by a factor f. This is clear from (2.85). After the expansion by a factor f the energy density du' in the new wavelength range λ' and $\lambda' + d\lambda'$ is decreased from the original energy density du due to two effects: (a) since the number of photons in a given volume that has increased due to the expansion of the universe remains the same, the photon density decreases by a factor f^3; ((b) since the energy of a photon is inversely proportional to its wavelength, its energy decreases by a factor f. Thus we get:

$$du' = (1/f^4)\,du = (8\pi hc/\lambda^5 f^4)\,d\lambda[\exp(hc/kT\lambda) - 1]^{-1}$$

$$= (8\pi hc/\lambda'^5)\,d\lambda'[\exp(hcf/kT\lambda') - 1]^{-1}. \quad (7.13)$$

This equation has the same form as (7.4) except that T has been replaced by T/f. It thus follows that freely expanding black-body radiation continues to be described by the Planck formula, but the temperature decreases in inverse proportion to R.

We can determine the neutrino temperature by considering the change in entropy as the universe expands. The entropy S at temperature T is proportional to $N_T T^3$, to a good approximation, where N_T is the effective number of species of particles in thermal equilibrium with threshold temperature below T. We have $N_T = N_1 N_2 N_3$, where N_1 is 1 if the particle does not have a distinct antiparticle, and 2 if it does; N_2 is the number of spin states of the particle; N_3 is a statistical mechanical factor which is $\frac{7}{8}$ or 1 according as to whether the particle is a fermion or a boson. In order to keep the total entropy constant, S must be proportional to R^{-3}, so that we have

$$N_T T^3 R^3 = \text{constant}. \quad (7.14)$$

As mentioned earlier, the neutrinos and antineutrinos went out of equilibrium with the rest of the contents of the universe before the annihilation of electrons and positrons (which occurred at approximately 5×10^9 K). Now according to the definition of N_T given above electrons and positrons have $N_T = \frac{7}{2}$, whereas photons have $N_T = 2$. Thus the total effective number of particles before and after the annihilation was

$$N_b = \frac{7}{2} + 2 = \frac{11}{2}; \quad N_a = 2. \quad (7.15)$$

From (7.14) it then follows that

$$\tfrac{11}{2}(T'R')^3 = 2(T''R'')^3, \quad (7.16)$$

where T', R' are values of T, R before annihilation, and T'', R'' the values afterwards. Thus

$$T''R''/T'R' = (\tfrac{11}{4})^{1/3} = 1.401. \quad (7.17)$$

This gives the increase in TR due to the heat produced by the annihilation.

The neutrino temperature T'_ν before the annihilation was the same as the photon temperature T'; from then on T'_ν just decreased like R^{-1}. Let the neutrino temperature afterwards be T''_ν. Thus

$$T''_\nu R'' = T'_\nu R' = T'R',\qquad(7.18)$$

from which, with the use of (7.17), it follows that

$$T''/T''_\nu = T''R''/T''_\nu R'' = T''R''/T'_\nu R' = T''R''/T'R'$$
$$= 1.401.\qquad(7.19)$$

Although the neutrinos go out of equilibrium quite early, they continue to make a significant contribution to the energy density. Remembering that the effective number of species N_T for neutrinos is $\frac{7}{2}$, and that the energy density is proportional to the fourth power of the temperature, the ratio of the densities of neutrinos to photons is:

$$u_\nu/u_\gamma = \tfrac{7}{4}(\tfrac{4}{11})^{4/3} = 0.4542.\qquad(7.20)$$

From (7.9) we see that the photon energy density u_γ can be written as follows:

$$u_\gamma = 7.5641 \times 10^{-15}T^4 \text{ erg cm}^{-3}.\qquad(7.21)$$

Thus the total energy density after the electrons and positrons have annihilated is

$$u = u_\gamma + u_\nu = 1.4542 u_\gamma = 1.100 \times 10^{-14}T^4 \text{ erg cm}^{-3}.\qquad(7.22)$$

The equivalent mass density is as follows:

$$\text{mass density} = u/c^2 = 1.22 \times 10^{-35}T^4 \text{ g cm}^{-3}.\qquad(7.23)$$

Given that the present temperature of the background radiation is of the order of 3 K, we see from (7.23) that the mass-energy density of this radiation is negligible in comparison to that of visible matter, which is of the order of 10^{-31} g cm^{-3}.

We have said earlier that the temperature decreases as R^{-1}. To examine this further consider the situation in which the rest masses of the particles are not necessarily negligible in comparison with their kinetic energies. Then the mass-energy density and the pressure are given as follows (we revert to ε):

$$\varepsilon = mn + \tfrac{3}{2}nkT + N'aT^4,\qquad(7.24a)$$

$$p = nkT + \tfrac{1}{3}N'aT^4,\qquad(7.24b)$$

where we envisage the contents to have a common temperature T, m being the mass of the massive particles (nucleons), k, a are the Boltzmann and Stefan constants, n is the number density of nucleons, and N' is related to the number of species of particles. The first terms in (7.24a) and (7.24b) give the non-relativistic contributions, the later ones give the relativistic terms. The number density n satisfies the following equation:

$$n(t)R^3(t) = \text{constant.} \tag{7.25}$$

This can be established from the baryon conservation law

$$J^\mu{}_{;\mu} = 0, \tag{7.26}$$

where the baryon current J^μ is given by $J^\mu = nu^\mu$, u^μ being the four-velocity. Equation (7.25) is then obtained from (7.26) with the use of (2.6a) and (2.105a)–(2.105d). We now substitute from (7.24a), (7.24b) and (7.25), into (2.112), which we write here again for convenience:

$$\dot{\varepsilon} + 3(p + \varepsilon)\dot{R}/R = 0. \tag{7.27}$$

The result of the substitution for $\dot{\varepsilon}$, $\dot{n}$, ε, p, into (7.27), is, after simplification, the following equation:

$$(\tfrac{1}{2} + N'\sigma)\dot{T}/T + (1 + N'\sigma)\dot{R}/R = 0, \tag{7.28}$$

where $\sigma = 4aT^3/3nk$. When $\sigma \gg 1$, (7.28) yields $TR = \text{constant}$ as a solution. In this case σ becomes a constant, since n varies as R^{-3}. This is termed a *hot universe*. To see what this implies, recall the number density of photons given by (7.11), which can be written as follows:

$$N = 20.28T^3 \text{ photons cm}^{-3} = 0.37(a/k)T^3 \text{ photons cm}^{-3}, \tag{7.29}$$

using the fact that $a = \pi^2 k^4/15c^3\hbar^3 = 7.5641 \times 10^{-15} \text{ erg cm}^{-3} \text{ K}^{-4}$, and $k = 1.38 \times 10^{-16} \text{ erg K}^{-1}$. Here $\hbar = h/2\pi$. From (7.29) and the definition of σ we see that

$$\sigma = 3.6 \, N/n. \tag{7.30}$$

Thus the condition $\sigma \gg 1$ implies that there are very many more photons and other relativistic particles than nucleons, so that radiation is unaffected by matter and after the decoupling of matter and radiation the temperature continues to drop like R^{-1}. The radiation maintains its black-body spectrum throughout the early universe as well as after the decoupling. We see from (7.25) and the decrease of T as R^{-1} that if the present number density of nucleons and the present temperature are respectively n_0 and T_0 and if these quantities have values n_1, T_1 respectively (the temperature being that of the background radiation), then the following relation obtains:

$$T_0 = (n_0/n_1)^{1/3}T_1. \tag{7.31}$$

If one can make a reasonable estimate of the nucleon number density at some early epoch, say when deuterium was just being formed (just below 10^9 K or so), one could predict the present temperature of the radiation from (7.31) and a knowledge of the present number density of nucleons. We will come back to this point later. Alternatively, one can use the present observed value 2.7 K of T_0 and an estimate of the present number density of nucleons to calculate the relation between T and n at any early epoch, and see what this implies for the abundances of the various nuclei. It is one of the successes of the standard model that the predictions of the abundances turn out to be in reasonable agreement with observed estimates.

7.5 Evolution of the mass-energy density

If we assume the early universe to be dominated by radiation, the equation of state is $p = \frac{1}{3}\varepsilon$, and (7.27) gives

$$\dot{R}/R = -\frac{1}{4}\dot{\varepsilon}/\varepsilon, \tag{7.32}$$

so that, with the use of (7.1) we get

$$\dot{\varepsilon} = -4(8\pi G/3)^{1/2}\varepsilon^{3/2}/c, \tag{7.33}$$

which can be integrated to give the following equation:

$$t = (3/32\pi G)^{1/2}\varepsilon^{-1/2}c + \text{constant}. \tag{7.34}$$

This relation, together with considerations of the previous section, leads to a thermal history of the early universe. This is done as follows. For any given range of temperatures, one determines the types of particles that are present in thermal equilibrium. One then determines the corresponding mass-energy density, assuming the particles to be relativistic. The temperature is given by Stefan's T^4 law. One then gets a relation between the time and the temperature with the use of (7.34). We will follow this procedure to provide a more quantitative description of the evolution of the early universe than that given at the beginning of this chapter. In this we follow mainly the accounts given by Weinberg (1972) and Bose (1980). We may repeat some parts of the qualitative account given earlier.

(i) 10^{12} K $> T > 5.5 \times 10^9$ K

Just below 10^{12} K the matter in the early universe consists of photons (γ), electron–positron pairs (e^-, e^+), electron- and muon-neutrinos and their antiparticles ($\nu_e, \nu_\mu, \bar{\nu}_e, \bar{\nu}_\mu$). There is also a small admixture of nucleons (neutrons and protons) and electrons – these will form the atoms of the later universe. Certain numbers of muons are also present to keep the neutrinos in thermal contact with other particles via weak interaction processes. The particles have a common temperature which is falling like R^{-1}. When the temperature goes below 10^{11} K or so, the neutrinos cease to be in thermal contact with the rest of the matter and radiation, but they continue to share a common temperature which drops like R^{-1}.

If the mixture of relativistic matter and radiation is considered to be an ideal gas, the number density $n_i(q) \, dq$ of particles of species i with momentum between q and $q + dq$ in thermal equilibrium is given as follows (Weinberg, 1972, Equation (15.6.3)):

$$n_i(q) \, dq = (4\pi/h^3)g_i q^2 \{\exp[(E_i(q) - \mu_i)/kT] \pm 1\} \, dq, \qquad (7.35)$$

where the positive sign applies for fermions and the negative for bosons. Since the particles are relativistic, the energy $E_i(q)$ of the ith particles with mass m_i is given by $c(q^2 + c^2 m_i^2)^{1/2}$, μ_i is the chemical potential of the ith species, g_i is the number of spin states, with $g = 1$ for neutrinos and antineutrinos, and $g = 2$ for photons, electrons, muons and their anti-particles.

The energy density for the ith species is given by

$$\varepsilon_i = \int_0^\infty E_i(q)n_i(q) \, dq, \qquad (7.36)$$

so that with the use of (7.35) one gets the following values for the photon and neutrino densities:

$$\varepsilon_\gamma = aT^4; \quad \varepsilon_\nu = \tfrac{7}{16}aT^4, \qquad (7.37)$$

where a is Stefan's constant mentioned earlier. The chemical potential for the photon is zero, so that for electrons and positrons it is equal and opposite, since chemical potential is conserved additively in reactions and $e^\pm$ pairs are produced from photons. However, in the range of temperatures under consideration, there are many more electron–positron pairs than unpaired electrons. Thus the number density of electrons is almost equal to that of positrons; since the corresponding chemical potentials are opposite it is reasonable to assume from (7.35) that both these chemical potentials vanish. Since the electrons are highly relativistic in the range of temperatures under considerations we can set $m_e \simeq 0$, and (7.36) yields the electron energy to be as follows:

$$\varepsilon_{e^-} = \tfrac{7}{8}aT^4. \qquad (7.38)$$

One can use these parameters to calculate the electron number density from (7.35) as follows:

$$n_{e^-} = \tfrac{3}{4}N, \tag{7.39}$$

where N is the photon number density given by (7.29). Since n, the nucleon density is nearly equal to the density of 'atomic' (unpaired) electrons, we see from (7.29), (7.30), (7.39) and the fact that $\sigma \gg 1$, that the electrons are predominantly the pair-produced ones. Adding the contributions of γ, ν_e, ν_μ, $\bar{\nu}_e$, $\bar{\nu}_\mu$, e^-, e^+, we get the total energy density to be as follows:

$$\varepsilon = \tfrac{9}{2}aT^4. \tag{7.40}$$

Putting this value of ε in (7.34) and inserting the values of a and G we get

$$t = 3.27 \times 10^{10}/T'^2 + \text{constant} = 1.09/T'^2 \text{ (s)} + \text{constant}, \tag{7.41}$$

where T' is the temperature measured in units of 10^{10} K. Thus the temperature takes 0.0108 s to drop from $T' = 10^2$ (that is, $T = 10^{12}$ K) to $T' = 10$ K ($T = 10^{11}$ K) and another 1.079 s to drop to $T' \simeq 1$, ($T = 10^{10}$ K). These values are roughly consistent with the 'first frame' time and temperature $t = 0.01$ s, $T = 10^{11}$ K, and 'third frame' $t = 1.1$ s, $T = 10^{10}$ K.

(ii) 5.5×10^9 K $> T > 10^9$ K

We have $m_e = 0.51$ MeV, so that the rest mass of an electron–positron pair is about 1.02 MeV. Thus the temperature at which electron–positron pairs are produced is given by $kT \simeq 1.02$ MeV, which yields, using the fact that $k^{-1} = 11\,605$ K eV^{-1}, a value of 1.1837×10^{10} K for the temperature at which pair production occurs. Thus at about 10^{10} K the electron–positron pairs start annihilating, and at the beginning of the present era these pairs become non-relativistic, so that (7.38) is no longer valid, and the behaviour $T \propto R^{-1}$ has to be modified. One can proceed by considering the entropy of particles in thermal equilibrium: electrons, positrons and photons. With the use of (7.35) one can work out the entropy in a volume R^3, as follows:

$$S = \frac{R^3}{T}(\varepsilon + p) = \frac{4a}{3}(RT)^3 \left\{ 1 + \frac{15}{2\pi^4} \int_0^\infty \frac{x^2(x^2 + 3y^2)}{y[\exp(y) + 1]}\, dx \right\}, \tag{7.42}$$

where $x = q/kT$, $y = E/kT$, $E = c(q^2 + c^2 m_e^2)^{1/2}$. Since the entropy S is constant, one can use (7.42) to determine how T changes with R. When the electrons are relativistic, we have $m_e \approx 0$, $x \approx y$, and the expression in the curly brackets becomes $(1 + \frac{7}{4})$; for non-relativistic electrons this factor is 1, so that $(RT)^3$ increases by a factor $\frac{11}{4}$. The ratio of the photon to neutrino temperatures, as we saw in (7.19), becomes $(\frac{11}{4})^{1/3} \approx 1.401$.

(iii) $T < 10^9$ K

The electron–positron pairs have annihilated completely and the particles in equilibrium are photons and the relatively small number of 'atomic' electrons and nucleons. The neutrinos have been decoupled for some time and are expanding freely. The corresponding temperatures and energy densities are worked out in (7.17)–(7.23), with the electron–nucleon densities negligible at the beginning of this era. From (7.20)–(7.22) we see that the energy density in the early stages of this era is

$$\varepsilon = [1 + \tfrac{7}{4}(\tfrac{4}{11})^{4/3}]aT^4 = 1.45aT^4. \tag{7.43}$$

Substituting in (7.34) we get

$$t = (15.5\pi Ga)^{-1/2}T^{-2}c + \text{constant} = 192T''^{-2}\,(s) + \text{constant}, \tag{7.44}$$

where T'' is measured in units of 10^9 K. By putting T'' equal to 1 and 0.1 respectively and subtracting, we see that it took about 5 h and 16.8 min for the temperature to drop from 10^9 K to 10^8 K. Equation (7.44) also gives the age of the universe at the time of recombination, that is, when electrons and protons combined to form hydrogen atoms at a temperature of about 4000 K, of about 4×10^5 years.

The onset of the matter-dominated era can be worked out as as follows. From (7.25) and the dropping of the photon temperature as R^{-1} we see the number density of nucleons satisfies

$$n/n_0 = (T/T_0)^3, \tag{7.45}$$

where n_0, T_0 are the present values of n, T. Thus the mass density of nucleons equals the density of radiation given by (7.43) at a temperature T_c which is as follows.

$$T_c = mn_0/(1.45aT_0^3). \tag{7.46}$$

If we take the present density of matter as 5×10^{-31} (this amounts to one-tenth of the critical density given by (3.9) if $H_0 \simeq 50$), then we get

$$T_c \simeq 2085 \text{ K}, \tag{7.47}$$

and the corresponding age of the universe from (7.44) is approximately

1.6×10^6 years. Thus the ages at which matter started becoming dominant and at which recombination occurred are of the same order of magnitude.

Thus in the early universe particles were highly relativistic most of the time and (7.32) and (7.34) are valid for that period; ε is found to be proportional to T^4 and (7.41) and (7.44) are obtained as the time–temperature relations, so that T decreases as R^{-1}. However, during the brief period of electron–positron annihilation the more complicated relation (7.42) obtains which can be written as

$$(RT)^3 F(T) = \text{constant}, \tag{7.48}$$

where $F(T)$ is a complicated function which becomes a constant both in the highly relativistic and fully non-relativistic regimes, yielding the usual behaviour $RT = \text{constant}$, but with different constants in the two regimes. The function $F(T)$ can be worked out by numerical methods; this becomes necessary if one wants to follow the details of the temperature drop which may be required for an analysis of nucleosynthesis.

Before we end this section we show explicitly how some of the figures mentioned at the beginning of this chapter are arrived at from the formalism given in this and the last two sections. The time and temperature for the first and third frames have already been mentioned in the paragraph containing (7.41). We are only concerned with the approximate derivation of the figures; a precise number containing several significant figures is not very meaningful in view of the uncertainties mentioned earlier, such as in the photon:baryon ratio.

If we assume that $t \ll 0.01$ s for some large value of T such as 10^{14} K, we can take the constant in (7.41) as negligibly small for our purpose. Then if we set $T' = 0.3$ K (which is the fourth frame temperature), we get $t = 109/9$ s $\simeq 12.1$ s. This is consistent with $t = 13$ s mentioned for the fourth frame, because we are just outside the range for which (7.41) is applicable, and t is a little higher than that given by (7.41). For the fifth frame (7.44) is just beginning to be applicable and for this frame we have $T'' = 1$, so that (again assuming the constant to be negligible), $t = 192$ s, which is consistent with t approximately 3 min given for the fifth frame. Similarly, for the sixth frame we put $T'' = 0.3$ K in (7.44) and get approximately 35 min for t. Also, when T is 4000 K at recombination, we get t as several hundred thousand years from (7.44), as mentioned towards the end of Section 7.1.

As an example of the energy density, from (7.23), taking T to be the third frame temperature of 10^{10} K, we get ε to be 1.22×10^5 cm^{-3}, which is consistent with the value mentioned for the third frame. Lastly, we give an example of the calculation of the Hubble time, which is the characteristic expansion time of the universe, given by $H^{-1} = R/\dot{R}$. From (7.32) and (7.33) we see that $R/\dot{R} = (8\pi G/3)^{-1/2}\varepsilon^{-1/2}$, which gives about 3 or 4 s as the third frame Hubble time, as mentioned.

7.6   Nucleosynthesis in the early universe

We have seen that in the early universe when the temperature was high enough neutrons and protons were separate and independent entities. In the present universe there are scarcely any free neutrons left; they form part of helium or heavier nuclei. In fact about 70–80% of the matter in the present universe is in the form of hydrogen, about 20–30% in the form of helium and a small percentage in the form of heavier nuclei. Any satisfactory theory of the early universe must explain the present observed abundances of the elements. One place in which nucleosynthesis can take place in the later universe, as mentioned earlier, is the centre of stars, where the temperature is of the order of a million Kelvin. Many of the heavier nuclei can indeed be produced here, as was shown in a famous paper by Burbidge, Burbidge, Fowler and Hoyle (1957). However, a simple calculation shows that the 20–30% helium that is observed today could not have been produced in the centre of stars. Indeed, the rate of energy release of our galaxy, for example, is about 0.2 erg g^{-1} s^{-1}. If the galaxy has been in existence for about 10^{10} years, this gives a total energy radiation of about 0.6×10^{17} erg per gram, or 0.375×10^{23} MeV per gram. Using the fact that a nucleon has mass 1.67×10^{-24} g, we see that this amounts to energy release of about 0.0625 MeV per nucleon, whereas hydrogen fusion into helium releases about 6 MeV per nucleon, so that only about 1% of the hydrogen in our galaxy could have been converted into helium.

In this section we will give an account of nucleosynthesis in the early universe. This is mainly based on Peebles (1971), Weinberg (1972), Schramm and Wagoner (1974) and Bose (1980).

The original suggestion that helium was synthesized in the early universe was made by Gamow, who developed a theory of nucleosynthesis with his collaborators in the 1940s. Although this theory was incomplete in some respects, there were useful insights and, in fact, a cosmic background radiation with temperature of 5 K was predicted in the 1940s! However, for various reasons this theory was not taken seriously. Gamow realized that helium synthesis was possible only during a brief period in the early universe (the first few minutes) and that for a sufficient amount of helium to be produced the density must have been very high. This leads to the picture of a hot and dense early universe, a picture which is essential in understanding nucleosynthesis in the early universe. One can start with the presently observed 2.7 K as the temperature of the remnant radiation and work backwards. This was done by Peebles (1971) and with other reasonable assumptions he obtained a helium abundance of about 25%. This is one of the conspicuous successes of the picture of an early universe that is hot and dense.

To work out the details one has to determine how the neutron–proton balance changes as the universe evolves; see if the rate of deuterium formation is sufficiently fast to ensure that nearly all the neutrons are used up; see if the reactions are fast enough to convert nearly all the deuterium into helium.

Neutrons and protons are converted into each other by the following weak processes:

$$n \leftrightarrow p + e^- + \bar{v}; \quad n + e^+ \leftrightarrow p + \bar{v}; \quad n + v \leftrightarrow p + e^-. \quad (7.49)$$

In the equilibrium condition as many neutrons are changing into protons as protons into neutrons. In the temperature range of interest the distribution of nucleons is given as follows, assuming they are non-relativistic (henceforth in the book we set $c = 1$ except for some specific cases):

$$n(q) \, dq = (8\pi/h^3) \exp[(\mu - m)/kT - q^2/2mkT]q^2 \, dq. \quad (7.50)$$

Here μ is the chemical potential of neutrons and protons, these being the same since the chemical potential is additively conserved, as noted earlier, and since leptons have zero chemical potential. In (7.50) m is the mass of the nucleon in units of energy, with $m_n - m_p \equiv Q = 1.293$ MeV. Integrating (7.50) between zero and infinity and taking the ratio of the cases for neutrons and protons respectively, we get:

$$n'/n'' = \exp(-Q/kT), \quad (7.51)$$

where n' and n'' denote the neutron and proton number densities respectively. Note that n', n'' become equal as T tends to infinity, or t tends to zero.

The number densities for v, $\bar{v}$, e^- and e^+ are given by (7.35) with zero chemical potential, with temperature T for $e^\pm$ and γ, and T_v for v, $\bar{v}$:

$$n_{e^-}(q) \, dq = n_{e^+}(q) \, dq = (8\pi/h^3)q^2 \, dq\{\exp[E_e(q)/kT] + 1\}^{-1}, \quad (7.52a)$$

$$n_v(q) \, dq = n_{\bar{v}}(q) \, dq = (4\pi/h^3)q^2 \, dq\{\exp[E_v(q)/kT_v] + 1\}^{-1}, \quad (7.52b)$$

where $E_e(q) = (q^2 + m_e^2)^{1/2}$ and $E_v(q) = q$, are the electron (or positron) and neutrino energies respectively. The rates of the reactions given in

E

(7.49) are given by the V–A theory of weak interactions (see, for example, Marshak, Riazuddin and Ryan, 1969), with the proviso that the Pauli exclusion principle decreases these rates by a factor corresponding to fraction of states unfilled, as follows:

$$1 - [\exp(E_e/kT) + 1]^{-1} = [1 + \exp(-E_e/kT)]^{-1}, \tag{7.53a}$$

$$1 - [\exp(E_v/kT_v) + 1]^{-1} = [1 + \exp(-E_v/kT_v)]^{-1}. \tag{7.53b}$$

Taking into account (7.52a), (7.52b), (7.53a) and (7.53b), the rates of the processes (7.49) per nucleon are given as follows:

$$\lambda(n + v \to p + e^-)$$

$$= A \int v_e E_e^2 q_v^2 \, dq_v [\exp(E_v/kT_v) + 1]^{-1} [1 + \exp(-E_e/kT)]^{-1}, \tag{7.54a}$$

$$\lambda(n + e^+ \to p + \bar{v})$$

$$= A \int E_v^2 q_e^2 \, dq_e [\exp(E_e/kT) + 1]^{-1} [1 + \exp(-E_v/kT_v)]^{-1}, \tag{7.54b}$$

$$\lambda(n \to p + e^- + \bar{v})$$

$$= A \int v_e E_v^2 E_e^2 \, dq_v [1 + \exp(-E_v/kT_v)]^{-1} [1 + \exp(-E_e/kT)]^{-1}, \tag{7.54c}$$

$$\lambda(p + e^- \to n + v)$$

$$= A \int E_v^2 q_e^2 \, dq_e [\exp(E_e/kT) + 1]^{-1} [1 + \exp(-E_v/kT_v)]^{-1}, \tag{7.54d}$$

$$\lambda(p + \bar{v} \to n + e^+)$$

$$= A \int v_e E_e^2 q_v^2 \, dq_v [\exp(E_v/kT_v) + 1]^{-1} [1 + \exp(-E_e/kT)]^{-1}, \tag{7.54e}$$

$$\lambda(p + e^- + \bar{v} \to n)$$

$$= A \int v_e E_e^2 q_v^2 \, dq_v [\exp(E_e/kT) + 1]^{-1} [\exp(E_v/kT_v) + 1]^{-1}. \tag{7.54f}$$

The constant A here is given as follows:

$$A = (g_V^2 + 3g_A^2)/2\pi^3\hbar^7, \tag{7.55}$$

with g_V and g_A being the vector and axial vector coupling constants of the nucleon, with the following values:

$$g_V = 1.407 \times 10^{-49} \text{ erg cm}^3; \quad g_A = -1.25g_V, \tag{7.56}$$

which correspond to a half-life of ~ 11 min for the decay of a free neutron. The lepton energies are related to Q as follows:

$$E_v + E_e = Q \quad \text{for} \quad n \leftrightarrow p + e^- + \bar{v}, \tag{7.57a}$$

$$E_v - E_e = Q \quad \text{for} \quad n + e^+ \leftrightarrow p + \bar{v}, \tag{7.57b}$$

$$E_e - E_v = Q \quad \text{for} \quad n + v \leftrightarrow p + e^-. \tag{7.57c}$$

In (7.54a)–(7.54f) v_e is the velocity of the electron given by q_e/E_e. These integrals are over lepton momenta that are consistent with (7.57a)–(7.57c). If these integrals are written over a common variable q $(=E_v = q_v)$ in (7.54a) and (7.54d) and as $-E_v$ in (7.54b), (7.54c), (7.54e) and (7.54f) and we also replace $q_e^2 \, dq_e$ with $v_e E_e^2 \, dE_e$, the total transition rates for $n \to p$ and $p \to n$ can be written as follows:

$$\lambda(n \to p) = \lambda(n \to p + e^- + \bar{v}) + \lambda(n + e^+ \to p + \bar{v}) + \lambda(n + v \to p + e^-)$$

$$= A \int \left[1 - \frac{m_e^2}{(q+Q)^2} \right]^{1/2} (q+Q)^2 q^2 \, dq [1 + \exp(q/kT_v)]^{-1}$$

$$\times \{1 + \exp[-(q+Q)/kT]\}^{-1}, \tag{7.58a}$$

$$\lambda(p \to n) = \lambda(p + e^- + \bar{v} \to n) + (p + \bar{v} \to n + e^+) + \lambda(p + e^- \to n + v)$$

$$= A \int \left[1 - \frac{m_e^2}{(q+Q)^2} \right]^{1/2} (q+Q)^2 q^2 \, dq [1 + \exp(-q/kT_v)]^{-1}$$

$$\times \{1 + \exp[(q+Q)/kT]\}^{-1}. \tag{7.58b}$$

Here T is the temperature of the electrons, photons and nucleons and T_v is the neutrino temperature; below about 10^{10} K, T and T_v are different and are given by (7.19). The integration in (7.58a) and (7.59b) ranges from $-\infty$ to $+\infty$ with a gap from $-Q - m_e$ to $-Q + m_e$. We are interested in the fractional abundance x given by

$$x = n'/(n' + n''), \tag{7.59}$$

whose evolution is given by the following equation:

$$dx/dt = -\lambda(n \to p)x + \lambda(p \to n)(1 - x). \tag{7.60}$$

In the limiting case when kT is much larger than Q, (7.58a) and (7.58b) yield the following approximations:

$$\lambda(p \to n) \simeq \lambda(n \to p) \simeq A \int q^4 \, dq [1 + \exp(-q/kT)]^{-1} [1 + \exp(q/kT)]^{-1}$$

$$= \tfrac{7}{15} \pi^4 A (kT)^5 = 0.36 T'^5 \text{ s}^{-1}, \tag{7.61}$$

Table 7.5 *Neutron fractional abundances as a function of time. (Taken from Peebles (1971).).*

T (K)	t (s)	$\lambda(\text{p} \rightarrow \text{n})$ (s^{-1})	$\lambda(\text{n} \rightarrow \text{p})$ (s^{-1})	x
10^{12}	0.00010	4.02×10^9	4.08×10^9	0.496
10^{11}	0.0109	3.9×10^4	4.6×10^4	0.462
2×10^{10}	0.273	9	19	0.330
10^{10}	1.102	0.19	0.83	0.238
10^9	182	0	0.00109	0.130
8×10^8	296	0	0.00108	0.116
6×10^8	535	0	0.00107	0.089

where, as in (7.41), T' is the temperature measured in units of 10^{10} K. We also have from (7.1), (7.40):

$$\dot{R}/R = (12\pi a G)^{1/2} T^2 = 0.46 T'^2 \text{ s}^{-1}. \tag{7.62}$$

From (7.61) and (7.62) we see that at $T' = 1$ ($T = 10^{10}$ K) a neutron is converting into a proton (and vice versa) at almost the same rate at which the universe is expanding. Thus at temperatures higher than 10^{10} K or so the processes (7.49) attain equilibrium and (7.50) is valid, and initially the neutron/proton numbers are nearly equal. Below 10^{10} K or so one has to integrate (7.58a), (7.58b), (7.59) and (7.60) numerically. This was done by Peebles (1971) and the results are set out in Table 7.5.

Helium synthesis involves essentially three steps. First, deuterium is produced (at a suitable temperature) directly from neutrons and protons. Next, two deuterium nuclei produce He^3 or H^3. The latter two nuclei then produce He^4, which is the stable helium isotope. The precise working out of helium synthesis is a complicated matter involving many equations. Such details have been considered by Peebles (1966) and by Wagoner, Fowler and Hoyle (1967). The reactions involved are many, such as:

$$\left.\begin{array}{l} \text{p} + \text{n} \leftrightarrow \text{D} + \gamma; \quad \text{D} + \text{D} \leftrightarrow \text{He}^3 + \text{n} \leftrightarrow \text{H}^3 + \text{p}; \\[4pt] \text{H}^3 + \text{D} \leftrightarrow \text{He}^4 + \text{n}; \quad \text{p} + \text{D} \leftrightarrow \text{He}^3 + \gamma; \quad \text{n} + \text{D} \leftrightarrow \text{H}^3 + \gamma; \\[4pt] \text{p} + \text{H}^3 \leftrightarrow \text{He}^4 + \gamma; \quad \text{n} + \text{He}^3 \leftrightarrow \text{He}^4 + \gamma; \quad \text{D} + \text{D} \leftrightarrow \text{He}^4 + \gamma. \end{array}\right\} \tag{7.63}$$

Reactions involving γs (photons) are radiative processes which usually take longer than other ones. Nucleosynthesis, when it begins, proceeds very quickly. The precise temperature at which it begins depends on the density, which can be extrapolated backwards from the present density, knowing the temperature of the background radiation. Peebles finds that nucleosynthesis begins at $T = 0.9 \times 10^9$ K if the present density is $\varepsilon_0 \simeq 7 \times 10^{-31}$ g cm^{-3}, or at $T = 1.1 \times 10^9$ K if it is $\varepsilon_0 \simeq 1.8 \times 10^{-29}$ g cm^{-3}. All processes which are relevant conserve the total number of nucleons. One result of nucleosynthesis is that the neutron:proton ratio is 'frozen' at the value it had just before nucleosynthesis began because once inside a nucleus a neutron cannot undergo beta decay. Before nucleosynthesis began, the ratio of neutrons to all nucleons is given by x (see (7.59)). After nucleosynthesis there are just free protons and He^4 nuclei. Thus the fraction of neutrons to all nucleons is just half the fraction of nucleons bound in He^4; this is the same as the abundance of helium by weight. It is found that a probable value (which comes out of the above calculations of x) when nucleosynthesis begins is 0.12. Thus the theory predicts about 24% for helium abundance, which is consistent with the observed value.

Appreciable amounts of elements heavier than helium cannot be produced in the early universe as there are no stable nuclei with five or eight nucleons, as mentioned earlier. As regards nuclei with seven nucleons, the Coulomb barrier (repulsion between the protons in different nuclei) in the reactions

$$\text{He}^4 + \text{H}^3 \to \text{Li}^7 + \gamma; \quad \text{He}^4 + \text{He}^3 \to \text{Be}^7 + \gamma, \qquad (7.64)$$

prevents these in comparison with

$$\text{p} + \text{H}^3 \to \text{He}^4 + \gamma; \quad \text{n} + \text{He}^3 \to \text{He}^4 + \gamma. \qquad (7.65)$$

He^4 has the highest binding energy by far of all nuclei with less than five nucleons, so effectively all the neutrons are used up in the formation of He^4.

7.7 Further remarks about helium and deuterium

We have seen earlier that the standard model predicts that the proportion of helium and deuterium present in the universe depends on the baryon: photon ratio. The helium abundance is higher for a greater number of baryons, while the deuterium abundance is correspondingly lower. The baryon:photon ratio is thus a crucial parameter in cosmology. As the cosmic background temperature is known fairly accurately, and as the photons in the present universe reside predominantly in the background radiation, the baryon:photon ratio can be worked out if one knows the matter density of the present universe, as the matter is predominantly in the form of baryons. Thus an accurate observational determination of the matter density, and of the relative abundances of helium and deuterium, can provide a useful test of the standard model.

To settle this question one has to examine if there are processes in the later universe which can create or destroy helium and deuterium. As we remarked earlier, significant amounts of helium could not have been produced in the later universe. One has to ask a similar question about deuterium. A brief discussion of deuterium production and destruction is in order here. In the Sun and such typical hydrogen burning main sequer ¹ stars deuterium is produced by weak interaction as follows:

$$p + p \rightarrow D + e^+ + \nu_e. \tag{7.66}$$

The deuterium thus produced is quickly transformed by the much faster reaction

$$p + D \rightarrow He^3 + \gamma. \tag{7.67}$$

Reactions (7.66) and (7.67) lead to a small equilibrium abundance of deuterium. The small amount of deuterium that is present in the interstellar medium and that is incorporated in stars soon disintegrates due to reactions such as (7.67). Thus any deuterium that existed when the galaxy was formed would be depleted by now. As mentioned earlier, deuterium is also created by the following radiative process:

$$p + n \rightarrow D + \gamma, \tag{7.68}$$

which is not prevented by the Coulomb barrier and involves no weak interaction. However, the free neutron that (7.68) requires is not usually present in astrophysical situations, except where there is very high energy involved such as in supernova explosions.

Another astrophysical situation in which deuterium can be created is in spallation reactions, mainly through the following reaction:

$$p + He^4 \rightarrow D + He^3. \tag{7.69}$$

This requires a centre-of-mass energy of 18.35 MeV, which is very high, because the binding energies of the product nuclei are somewhat less than those of the initial ones. In (7.69), for example, a part of this energy is used up in extracting the neutron from the He^4 nucleus. Such high energies sometimes exist in cosmic ray protons.

In astrophysical settings deuterium can be readily destroyed by the following reactions:

$$D + D \rightarrow n + He^3; \quad D + D \rightarrow H^3 + p; \quad n + D \rightarrow H^3 + \gamma, \tag{7.70}$$

if either the neutron or deuterium concentration is high. Thus to produce and preserve deuterium one needs high energy and low density.

The abundance of deuterium is usually specified by D/H, the ratio of deuterium and hydrogen nuclei in a small volume. This ratio is different in different astrophysical and terrestrial situations. In sea water, for example, where deuterium occurs as HDO (heavy water; obtained by replacing a hydrogen atom in H_2O by deuterium), the ratio is 150 ppm (parts per million), which is somewhat higher than the average for all situations. The proportion of deuterium in carbonaceous meteorites is similarly high. The high proportion of deuterium in sea water is explained by the fact that due to chemical fractionation, in the formation of water D is preferred to H; the larger mass of D allows for different chemical and nuclear properties. On the other hand, in the outer regions of the Sun D/H is only about 4 ppm. This is because reactions such as $D + p \rightarrow He^3 + \gamma$, destroy deuterium in the Sun. In the interstellar gas near the Sun D/H is about 14 ppm. In the interstellar gas deuterium is detected through deuterated molecules such as CH_3D (deuterated methane) and DCN (deuterium cyanide). For example, it was found that in the Orion nebula the DCN/HCN ratio was about 40 times the terrestrial D/H ratio (Jefferts, Penzias and Wilson, 1973; Wilson, Penzias, Jefferts and Solomon, 1973); this is again due to chemical fractionation which favours DCN formation over HCN. The deuterium in interstellar material is detected by its 91.6 cm hyperfine line (the equivalent of the well-known 21 cm hydrogen line). The possibility of deuterium production in super-nova explosions has also been considered (see Schramm and Wagoner (1974) for references on this), but it is found that these explosions are much more efficient at producing other light elements such as Li^7, Be^9 and B^{11} than D. In Table 7.6 we set out the observed abundances of deuterium in various situations; although some of these may be out of date the table nevertheless incorporates some essential points.

As the Jovian CH_3D estimate requires determination of the CH_4 abundance there was some uncertainty about this measurement (Beer and Taylor, 1973). The Voyager infrared experiment enabled a simultaneous determination of CH_4/CH_3D mixing ratio and Kunde *et al.* (1982) then derived the D/H ratio from Jovian CH_3D as 22 and 46 ppm (see Gautier and Owen (1983)). This is not inconsistent with the 1973 estimate given by Beer and Taylor (see Table 7.6).

The question arises as to what extent the helium and deuterium abundances found in the present universe represent these abundances in the primordial universe soon after nucleosynthesis. As we have seen, deuterium can be created to a small extent and destroyed more readily in the later universe. For helium a minor component of the abundance currently observed can be produced in stars and injected into the

Table 7.6 *Observed ratio of deuterium to hydrogen atoms. (Reproduced from Schramm and Wagoner (1974) with minor omissions.)*

	Location	$(D/H) \times 10^6$ (ppm)	Observer
Solar system	Earth (HDO)	150	Friedman et al, 1964
	Meteorites (HDO)	130–200	Boato, 1954
	Jupiter (CH_3D)	28–75	Beer and Taylor, 1973
	Jupiter (HD)	21 ± 4	Trauger et al, 1973
	Present Sun	<4	Grevesse, 1970
	Primordial Sun:		
	From He^3 in gas-rich meteorites	10–30	Black, 1971, 1972
	From He^3 in solar wind	<50	Geiss and Reeves, 1972
	From He^3 in solar prominences	<60	Hall, unpublished
Interstellar medium	Cassiopeia A (91.6 cm line)	<70	Weinreb, 1962
	Sagittarius A (1.6 cm line)	<350	Cesarsky, Moffet and Pasachoff, 1973,
			Pasachoff and Cesarsky, 1974
	β Centauri	14 ± 2	Rogerson and York, 1973

Table 7.7 *Helium abundances.* (*Taken from Gautier and Owen* (1983) *with some omissions and a minor change.*)

Determination	Y	Reference
Jupiter (Voyager IRIS)	$0.15 < Y < 0.24$	Gautier *et al*, 1981
Saturn:		
Pioneer 11	0.18 ± 0.05	Orton and Ingersoll 1980
Voyager IRIS	~ 0.14	Conrath, Gautier and Hornstein 1982
Solar:		
Helium emission lines	0.28 ± 0.05	Heasley and Milkey 1978
Cosmic rays	0.20 ± 0.04	Lambert, 1967
Standard interior models	0.22	Iben, 1969; Bahcall *et al*, 1973, 1980; Ulrich and Rood, 1973; Mazzitelli, 1979
Primordial		
best estimate from several		
results	0.23 ± 0.01	Pagel, 1984

interstellar medium by supernova explosions and stellar winds. The giant planets like Jupiter and Saturn, because of their low exospheric temperatures and large masses, provide environments in which the elements are more or less in their primordial form, almost undisturbed for about 4.55 billion years since these planets were formed. Even the lightest elements have not escaped from the atmospheres of these planets since their inception. The Jovian helium abundance has been determined by Voyager. The hydrogen/helium mixing ratio can be found in many different areas of Jupiter and one finds a mass ratio Y (Y = mass of helium/mass of all nuclei) of 0.19 ± 0.05 by one method, and $Y = 0.21 \pm 0.06$ by another. Combining the two methods one gets (Gautier and Owen, 1983):

$$0.15 < Y < 0.24. \tag{7.71}$$

Table 7.7 summarizes a representative set of observations of helium abundance. The low value of Y for Saturn is probably due to the phenomenon of differentiation of helium from hydrogen (Smoluchowski, 1967) that may have depleted the amount of helium in the Saturn atmosphere. Presumably this phenomenon has not begun in Jupiter.

Gautier and Owen (1983) find that the primordial abundance of deuterium must have been reduced (that is, the deuterium must have been destroyed) by a factor of between 5 and 16 between the time of the primordial nucleosynthesis and the origin of the solar system 4.55 billion years ago. This is seen as follows. Since helium and deuterium were synthesized at the same time, Y and $X(D)$ (the deuterium mass fraction which is approximately 1.5 times D/H – the exact multiple depending on Y) have a certain dependence on η, the ratio of baryon to photon number densities. The uncertainty in Y_p (the primordial value of Y) is found to be: $0.22 < Y_p < 0.24$, which corresponds to the following uncertainty in η: $0.35 \times 10^{-11} < \eta < 2 \times 10^{-10}$. The corresponding abundance of primordial deuterium turns out to be $3.4 \times 10^{-4} < (X(D))_p < 11.6 \times 10^{-4}$. From the Jovian deuterium abundance one gets the upper limit $X(D) < 7 \times 10^{-5}$. This leads to the discrepancy cited at the beginning of this paragraph. This analysis seems to imply that either deuterium is destroyed more efficiently than hitherto assumed, or that the standard model needs some modification. Whether this claim made by Gautier and Owen is valid is not clear, but the above analysis does emphasize the need to look very carefully into the question of helium and deuterium abundances and their relation with the baryon to photon number density ratio, both observationally and theoretically. There are some other assumptions made in this analysis which we have not mentioned; one of these is the assumption that there are three different kinds of neutrinos. The reader is referred to Gautier and Owen (1983) for more details.

As noted earlier, different amounts of nuclei are created in the early universe according to different assumptions of the baryon:photon ratio, which, in turn, depends on the present mass density of the universe. Thus different values of the present mass density give different abundances. Figure 7.3 depicts this dependence of the abundances on the mass density, as given by Schramm and Wagoner (1974). It is interesting that the He^4 abundance is almost constant, that is, it is not at all sensitive to the value of the present mass density. By contrast, the abundance of D is strongly dependent on the mass density.

7.8   Neutrino types and masses

We end this chapter with a brief discussion of neutrino types and masses and the cosmological implications of these. We saw earlier that the temperature depends on the types of particles that were in thermal equilibrium with the photons in the early universe. In our earlier analysis we did not adequately take into account the fact that there are different types of neutrinos. Two types, the electron- and muon-neutrinos, are definitely known. There may be a third type associated with the heavier tau-lepton, which was discovered relatively recently. If there are three or

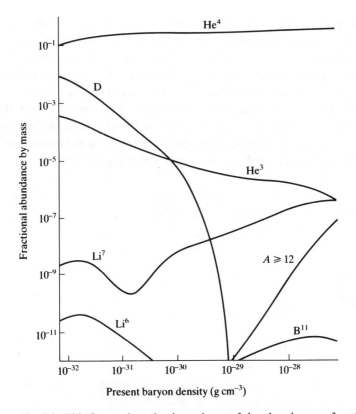

Fig. 7.3. This figure gives the dependence of the abundances of various nuclei on the present value of the mass density, which is not precisely known. The curve marked $A \geqslant 12$ refers to nuclei with baryon number greater than or equal to 12.

more kinds of neutrinos, it can be shown that this results in faster expansion in the early universe, so that more He^4 is produced. However, like the mass density, the He^4 abundance is not so sensitively dependent on neutrino types so that many more types than are known at present can be accommodated without seriously violating the observed He^4 abundance. However, it is quite a different matter with D abundance, which is highly sensitive to the number of neutrino types. It would be very difficult to reconcile the observed D abundance if the neutrino types were five or six in number. However, the latter situation would be saved somewhat if the neutrinos had mass, as has been indicated recently. Massive neutrinos have much less effect on the expansion rate and nucleosynthesis in the early universe. Another consequence of massive neutrinos is that the 'background' neutrinos then might contribute enough mass to the present mass density to make it above the critical density. The present indications are that neutrino masses cannot be more than a few electron volts. A recent analysis of the neutrino arrival time from the supernova in the Large Magellanic Cloud (Hirata *et al*, 1987; Adams, 1988) shows that there is a 90% probability of the neutrino mass being less than 5 eV and 99% probability of it being less than 10 eV. A great deal of theoretical and observational work has to be done to clarify this question. We refer the interested reader to the papers cited, and Tayler (1983), Schramm (1982) and Bahcall and Haxton (1989).

8

The very early universe and inflation

8.1 Introduction

As is clear from the discussion so far in this book, the standard big bang model incorporates three important observations about the universe. These are firstly the expansion of the universe discovered by Hubble in the 1930s, the discovery of the microwave background radiation by Penzias and Wilson and its confirmation by other observers, and thirdly the prediction of the abundances of various nuclei on the basis of nucleosynthesis in the early universe, particularly the abundances of He^4 and deuterium, which appear to conform reasonably with observations. As is also clear from the earlier discussions, much theoretical and observational work remains to be done to clarify these questions further.

As mentioned in Chapter 1, some glaring puzzles do remain, such as the horizon problem. The puzzle here is: how is the universe so homogeneous and isotropic to such vast distances, extending to regions which could not have communicated with each other during the early eras? This problem is illustrated in Fig. 8.1. Another puzzle is why the density parameter Ω (the ratio of the energy density of the universe to the critical density – see discussion following (1.4)) is so near unity. If the present value of Ω lying between 0.1 and 2 is extrapolated to near the big bang we get the following orders of magnitude:

$$|\Omega(1\text{ s}) - 1| = O(10^{-16}), \tag{8.1a}$$

$$|\Omega(10^{-43}\text{ s}) - 1| = O(10^{-60}). \tag{8.1b}$$

These extremely small numbers seem difficult to explain. The third problem is the smoothness problem, which is to explain the origin and nature of the primordial density perturbations which result in the 'lumpiness', that is, the presence of galaxies and the structure of the observable universe. The inflationary models, of which the original one was propounded by Guth (1981), attempt to explain these puzzles.

8.2   Inflationary models – qualitative discussion

In this section we shall give a qualitative description of inflationary models; this will be followed by some quantitative accounts. However, it will not be possible to explain all aspects quantitatively. Some aspects involve fairly technical questions of particle physics and in particular Grand Unified Theories, which are beyond the scope of this book. Our treatment of inflationary models is by no means exhaustive; our intention is to point out the essential features.

As mentioned earlier, at high energies, according to the Glashow–Salam–Weinberg unified electroweak theory, electromagnetic and weak interactions behave in a similar manner, and consequently there is a phase transition in the early universe associated with this at a critical temperature of about 3×10^{15} K. The Grand Unified Theories attempt to find a unified description of all three of the fundamental interactions, namely, electromagnetic, weak and strong interactions. Grand Unified Theories predict that there is a phase transition in the universe at a critical

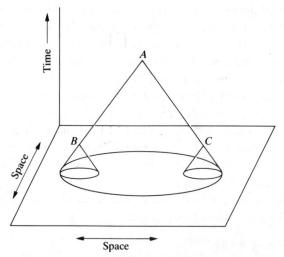

Fig. 8.1. This diagram illustrates the horizon problem. The point *A* represents our present space-time position, one space dimension being suppressed in this diagram. The points *B* and *C* represent events at a much earlier epoch, lying in opposite spatial direction from us, but lying in our past light cone. The plane at the bottom represents the instant $t = 0$, the big bang. The past light cones of *B* and *C* have no intersection, so these two events could not have had any causal connection. How is it that radiation received from these two points (the cosmic background radiation) are at the same temperature?

temperature of about 10^{27} K, above which there was a symmetry among the three interactions. Consider again the analogy with the freezing of water. In the liquid state there is rotational symmetry at any point in the body of the water; this symmetry is lost, or 'broken' when ice is formed, as ice crystals have certain preferential directions. Secondly, the liquids in different portions begin to freeze independently of each other with different crystal axes, so that when the whole body of the liquid is frozen certain defects remain at the boundaries of the different portions. In a similar manner in the early universe above 10^{27} K or so the symmetry among the three interactions was manifest, and below this temperature this symmetry was broken. Now in water the manner in which the rotational symmetry is broken in different portions can be characterized by parameters which describe the orientation of the ice-crystal axes. Thus these parameters take different values in different portions of the liquid as it freezes, that is, as the symmetry is broken. In a similar way, the manner in which the manifest symmetry among the three interactions is broken can be characterized by the acquiring of certain non-zero values of parameters known as Higgs fields; this is referred to as *spontaneous symmetry breaking*. The symmetry is manifest when the Higgs fields have the value zero; it is spontaneously broken whenever at least one of the Higgs fields becomes non-zero. Just as in the case of the freezing of water, certain defects remain at the boundaries of different regions in which the symmetry is broken in different ways, that is, by the acquiring of different sets of values for the Higgs fields. There are point-like defects which correspond to magnetic monopoles, and two dimensional defects called domain walls. A region in which the symmetry is broken in a particular manner could not have been significantly larger than the horizon distance at that time, so one can work out the minimum number of defects that must have occurred during the phase transition. The defects are expected to be very stable and massive. For example, it turns out that monopoles are about 10^{16} times as massive as a proton. The result is that there would be so many defects that the mass density would accelerate the subsequent evolution of the universe, so that the 3 K background radiation would be reached only a few tens of thousands of years after the big bang instead of ten billion years. Thus this prediction of Grand Unified Theories seems to conflict seriously with the standard model.

None of the successes of the standard model are affected by the inflationary models, because after the first 10^{-34} s or so, the two models are exactly the same as far as our observable universe is concerned. The original inflationary model put forward by Guth in 1981 had serious drawbacks, as mentioned in Chapter 1. We shall be concerned with the 'new inflation' put forward independently by Linde (1982) and Albrecht and Steinhardt (1982). For simplicity we consider a single Higgs field which we take to be a scalar field ϕ. The possible forms of the potential energy corresponding to this field are indicated in Figs. 8.2 and 8.3.

Consider some properties of the potential as depicted in Figs. 8.2 and 8.3. The potential has stationary points at $\phi = 0$ and $\phi = \sigma$. At these points the system can be in equilibrium. The states which are stationary states of the potential can be referred to as 'vacuum' states. Consider Fig. 8.2 first. The energy of the stationary state at $\phi = 0$ is higher than that at $\phi = \sigma$. There might be a situation in which the system is 'trapped' in

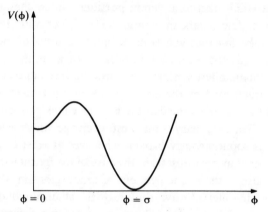

Fig. 8.2. One of the possible forms of the potential for the scalar field given by Equation (8.16a)

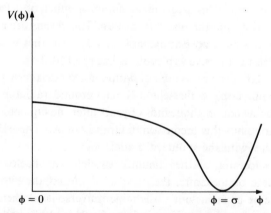

Fig. 8.3. Another possible form for the potential of the scalar field given by Equation (8.16b).

the stationary state at $\phi = 0$ and cannot make the transition to the stationary state at $\phi = \sigma$, because of the potential barrier, even though $\phi = \sigma$ has a lower energy. In this situation the state $\phi = 0$ is referred to as a 'false vacuum', while $\phi = \sigma$ is the 'true vacuum'. What is the relevance of this to the very early universe and inflation?

We assume that the very early universe had regions that were hotter than 10^{27} K and were expanding. The symmetry among the interactions was manifest and the Higgs field, represented by ϕ here, was zero. One can look upon this situation as the thermal fluctuations driving the Higgs field to the equilibrium value zero. As the expansion caused the temperature to fall, below the critical temperature it would be thermodynamically more favourable for the Higgs field to acquire a non-zero value. However, for some values of the parameters in Grand Unified Theories the phase transition occurs very slowly compared to the cooling rate. This can cause the temperature to fall well below 10^{27}, the critical temperature, but the Higgs field to remain zero. This is akin to the phenomenon of supercooling; for example, water can be supercooled to 20° below freezing. This is the situation of the false vacuum mentioned above, in which the Higgs field remains zero although it is energetically more favourable to go to the state $\phi = \sigma$ (that is, the energy in the state $\phi = \sigma$ is lower than that in $\phi = 0$). It turns out that this situation causes the region to cool down considerably and also have a very high rate of expansion. The situation, depicted in Fig. 8.2, however, leads to difficulties, which are avoided in that depicted in Fig. 8.3, so we shall follow the rest of the development in the latter situation.

Before considering a more quantitative description of inflationary models, it may be useful to give an idea of the overall effect of these models on the standard model. This is given in Fig. 8.4, which is taken from Turner (1985). The inflationary models incorporate all the predictions of the standard model for the observable universe, because for the latter the inflationary models have the same behaviour after $t = 10^{-32}$ s or so. From about 10^{-34} to 10^{-32} s or so, the inflationary models are radically different. A region of the universe underwent accelerated, exponential expansion, as well as cooling. After this period of expansion and cooling it was reheated to just below the critical temperature. After this the story is the same as the standard model, the important difference being, the initial region was within a horizon distance and had time to homogenize and have the same temperature, etc., and after the inflation the entire observable universe can lie within such a region, so that the horizon problem does not arise. Let us see this, still qualitatively, in more detail keeping in mind the Higgs potential of Fig. 8.3.

Consider a region of the very early universe which was hotter than 10^{27} K or so. We will consider the evolution of this region in the inflationary model. The reason this evolution is different from the standard model is due to the presence of the Higgs field, which describes the phase transition that the universe undergoes at that temperature, as mentioned earlier. The presence of the Higgs field radically alters the evolution of the scale factor R in the very early epochs, as depicted in

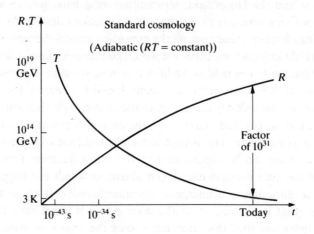

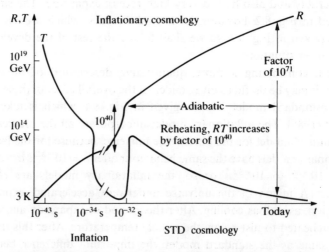

Fig. 8.4. This figure depicts the evolution of the scale factor R and temperature T of the universe in the standard model and in the inflationary models. The standard model is always adiabatic ($RT \approx$ constant), except for minor deviations when particle–antiparticle pairs annihilate, whereas inflationary models undergo a highly non-adiabatic event (at 10^{-34} s or so), after which it is adiabatic. (Turner, 1985.)

Fig. 8.4, the evolution of R being the same as in the standard model after 10^{-32} s or so. We will write down these equations in the next section. Here we describe this evolution qualitatively with the Higgs potential being given by Fig. 8.3. As mentioned earlier, above 10^{27} K thermal fluctuations drive the equilibrium value of the Higgs field to zero and the symmetry is manifest. As the temperature falls the system undergoes a phase transition with at least one of the Higgs fields acquiring a non-zero value (here we consider only one), resulting in a broken symmetry phase. However, for certain values of the parameters, which we assume to be the case, the rate of the phase transition is very slow compared with the rate of cooling. This causes the system to supercool to a negligible temperature with the Higgs field remaining at zero (this corresponds to the 'well' in the curve marked T in the lower figure in Fig. 8.4), resulting in a 'false vacuum'. Now quantum fluctuations or small residual thermal fluctuations cause the Higgs field to deviate from zero. Unlike the situation depicted in Fig. 8.2, in Fig. 8.3 there is no energy barrier, so the Higgs field begins to increase steadily. The rate of increase is, in fact, like the speed of a ball which was perched on top of the potential curve in Fig. 8.3 (at $\phi = 0$) and which starts to roll down; at first the speed is very slow, increasing gradually until it has high speed in the steeper portions, and finally it oscillates back and forth when it reaches the bottom of the well. In the flatter portions, as we shall see more clearly in the next section, the region undergoes accelerated expansion, doubling in diameter every 10^{-34} s or so. When the value of the Higgs field reaches the steeper parts of the potential curve, the expansion ceases to accelerate. An expansion factor of 10^{50} or more can be achieved in this manner for the region under consideration.

The picture given above is a simplified one. As mentioned earlier there can be many different broken-symmetry states (depending on the non-zero values acquired by the Higgs fields), just as there are many different possible crystal axes during the freezing of a liquid. Thus different regions in the very early universe would acquire different broken-symmetry states, each region being roughly of the size of the horizon distance at the time. The horizon distance at time t is approximately ct, the distance travelled by light in time t; thus at $t = 10^{-34}$ s the horizon distance is about 10^{-24} cm. Once a domain was formed with a particular set of non-zero values of the Higgs fields, it would gradually attain one of the stable broken-symmetry states and inflate by a factor of 10^{50} or so. Thus after inflation the size of such a domain would be approximately 10^{26} cm. At that epoch the entire observable universe would measure only 10 cm or so, so it would easily fit well within a single domain. Since the observable universe lay within a region which, in turn, started from a region contained in a horizon distance, it would have had time to homogenize and attain a uniform temperature. This then solves the horizon problem.

Because of the enormous inflation, any particle with a certain density that may have been present before the inflation, would have its density reduced to almost zero after the inflation. Most of the energy density would be incorporated in the Higgs field after the inflation. After the Higgs field evolves away from the flatter portion of the curve in Fig. 8.3 and goes down the steep slope and starts oscillating back and forth near the true vacuum at $\phi = \sigma$, we have the situation that corresponds in quantum field theory to a high density of Higgs particles (recall a high level of energy for a harmonic oscillator corresponds to a larger number of 'excitations' of the electromagnetic field, that is, a large number of photons). The Higgs particles would be unstable and would undergo decays into lighter particles, and the system would rapidly attain the condition of a hot gas of elementary particles in equilibrium, akin to the initial condition assumed in the standard model. The system would be reheated to a temperature of about 2–10 times lower than the phase transition temperature of 10^{27} K. The story after this is the same as the standard model, so that the successes of the standard model are maintained.

Several points and questions remain in the above description, which we will deal with at the end of this chapter. Firstly, how is the monopole problem solved by this model? One of the problems of the standard model that Grand Unified Theories purport to solve is the problem of baryon asymmetry, that is, why do we see matter rather than antimatter in the present universe? In other words, when in the last chapter we spoke of a small 'contamination' of neutrons and protons, one can ask why there was not a contamination of antineutrons and antiprotons instead. Thirdly, as a matter of interest, what was wrong with the original model put forward by Guth (1981)? Lastly, do any problems remain in the new inflationary model as described above? In other words, is the new inflationary model able to solve all the problems of the standard model mentioned earlier and not throw up problems of its own, that is, is it self-consistent and in accord with present observations?

8.3   Inflationary models – quantitative description

As mentioned earlier, it is not possible to give here the technical details from particle physics and quantum field theory. Secondly, even in the classical and semi-classical treatments, suitable exact solutions are not known so that even when we have the equations a certain amount of qualitative analysis is necessary.

Recall Einstein's equations (2.22) (with $c = 1$):

$$R_{\mu\nu} - \tfrac{1}{2}g_{\mu\nu}R = T_{\mu\nu}.$$

Here $T_{\mu\nu}$ represents the energy–momentum tensor. For a perfect fluid this is given by (2.23). However, although the latter case suffices for the standard mode, in general, one has to consider the contributions to $T_{\mu\nu}$ from all possible fields. For example, when there is an electromagnetic field present (this is not relevant in the cosmological context), one has to add the following contribution to the energy–momentum tensor:

$$T_{\mu\nu}^{(\text{em})} = (4\pi)^{-1}(-F_\mu{}^\alpha F_{\nu\alpha} + \tfrac{1}{4}g_{\mu\nu}F_{\alpha\beta}F^{\alpha\beta}), \tag{8.2}$$

where the electromagnetic field tensor $F_{\mu\nu}$ is given in terms of the four-potential A_μ as follows: $F_{\mu\nu} = A_{\mu,\nu} - A_{\nu,\mu}$.

As mentioned earlier, the phase transition of the very early universe can be described by introducing a scalar Higgs field ϕ into the theory. One way to do this is to add an additional energy–momentum tensor $T'_{\mu\nu}$, due to the Higgs field, to the existing energy–momentum tensor on the right hand side of Einstein's equations (2.22). The form of this additional energy–momentum tensor is suggested by the Lagrangian of a scalar field, which is as follows (V is a suitable potential):

$$L = \tfrac{1}{2}\partial_\mu\phi\partial^\mu\phi - V(\phi). \tag{8.3}$$

It is well known (see, for example, Bogoliubov and Shirkov (1983, p. 17)) that for a scalar field the energy–momentum tensor associated with a Lagrangian L is given by

$$T_{\mu\nu} = \frac{\partial L}{\partial\phi^{,\nu}}\,\phi_{,\nu} - g_{\mu\nu}L. \tag{8.4}$$

For L given by (8.3) this gives

$$T'_{\mu\nu} = \partial_\mu\phi\,\partial_\nu\phi - g_{\mu\nu}L = \partial_\mu\phi\,\partial_\nu\phi - g_{\mu\nu}[\tfrac{1}{2}\partial_\sigma\phi\,\partial^\sigma\phi - V(\phi)]. \tag{8.5}$$

The energy–momentum tensor for a perfect fluid given by (2.23) for the comoving cosmological fluid can be written in the following form:

$$T_\mu{}^\nu = \text{diag}(\varepsilon, -p, -p, -p), \tag{8.6}$$

where the tensor is written in matrix form with diagonal elements, other elements being zero. We now assume that the scalar field, ϕ, depends on the time t only, and if we write the tensor $T'_{\mu\nu}$ in the form (8.6), that is,

$$T'^\nu_\mu = \text{diag}(\varepsilon', -p', -p', -p'), \qquad (8.7)$$

we find from (8.5) the following relations for ε', p':

$$\varepsilon' = \tfrac{1}{2}\dot\phi^2 + V(\phi); \quad p' = \tfrac{1}{2}\dot\phi^2 - V(\phi), \quad \dot\phi \equiv \partial\phi/\partial t. \qquad (8.8)$$

Thus the modified Einstein equations in the cosmological situation with the Higgs field ϕ are obtained by simply replacing ε by $\varepsilon + \varepsilon'$, and p by $p + p'$, with ε', p' given by (8.8).

There are some basic assumptions in this analysis which we must clarify. Firstly we are dealing essentially with a region which is within the horizon distance at the time under consideration. This region, according to the above scenario, undergoes rapid expansion, cooling, etc., more or less independent of the rest of the universe. Yet we are using for this region the Robertson–Walker metric which is derived under the assumption that the entire space is homogeneous and isotropic. We are thus using the assumption here that the total space-time behaves in such a manner that the Robertson–Walker form of the metric is justified locally. Secondly, we are ignoring the spatial variation of ϕ, so that throughout the region ϕ takes a uniform value. This assumption leads to the relatively simple Equations (8.8). A third assumption, which is not a serious restriction, is that in (8.8) we have used the $k = 0$ form of the Robertson–Walker metric, that is, the form which has flat spatial geometry. We will do this throughout this chapter. Thus the Einstein equations are now given by (with $c = 1$):

$$(\dot R/R)^2 \equiv H^2 = (8\pi G/3)(\varepsilon + \varepsilon'), \qquad (8.9a)$$

$$2\ddot R/R + H^2 = -8\pi G(p + p'), \qquad (8.9b)$$

where ε', p' are given by (8.8) in terms of ϕ.

Consider now the situation in the very early universe when the temperature is higher than 10^{27} K. As mentioned earlier $\phi = 0$ so that from (8.8) we see that ε' has the constant value $V(0)$. On the other hand, if we assume that the equation of state is that of radiation, we see that R behaves like $t^{1/2}$ (see (3.47)) while ε behaves like t^{-2} (see (3.40)). Thus in (8.9a) the ε dominates the right hand side, so the evolution of R is as if the ϕ term did not exist, that is, ε decreases like t^{-2} as t increases. Since the evolution of R is faster than the phase transition (for some set of parameters of Grand Unified Theories), ϕ remains at the value zero while the temperature goes below the critical. The ε term on the right hand side of (8.9a) becomes much less than ε', that is, $V(0)$, so that the evolution of

R is given by

$$(\dot{R}/R)^2 = (8\pi G/3)V(0), \tag{8.10}$$

which has the solution

$$R = \exp(\zeta t), \quad \zeta^2 = (8\pi G/3)V(0), \tag{8.11}$$

provided $V(0)$ is positive, which is the case, for example, in Figs. 8.2 and 8.3. Thus the scale factor R undergoes exponential expansion. This behaviour is, in fact, that of de Sitter space, for which we give a little digression.

In (5.2a) and (5.2b) if we set $\varepsilon = p = k = 0$, we get (with $c = 1$):

$$(\dot{R}/R)^2 = \tfrac{1}{3}\Lambda, \tag{8.12a}$$

$$2\ddot{R}/R + (\dot{R}/R)^2 = \Lambda. \tag{8.12b}$$

It is readily verified that (8.12a) and (8.12b) are satisfied by

$$R = \exp(\tfrac{1}{3}\Lambda)^{1/2}t, \tag{8.13}$$

assuming that the cosmological constant is positive. The model given by (8.13) (with $k = 0$) is called the *de Sitter universe*, which is empty, has a positive cosmological constant, and has a non-trivial scale factor given by (8.13). Sometimes it is said that the de Sitter space represents 'motion without matter' as opposed to the Einstein universe (see (5.3)), which represents 'matter without motion'. Equation (8.13) gives the same behaviour as (8.11) and is also the form of the steady state universe, as mentioned earlier, which was put forward originally by Bondi and Gold (1948) and by Hoyle (1948). The latter is maintained at a steady state by the continuous creation of matter, the amount of which is cosmologically significant but negligible by terrestrial standards, so that no experiment on the conservation of mass is violated. Observations of the background radiation and others, however, contradict the steady state theory. It is curious that in the inflationary models one has to consider again a similar exponential metric (8.11), albeit for a very short period in the history of the universe.

When the energy–momentum tensor $T_{\mu\nu}$ of the cosmological fluid can be neglected in comparison with the energy–momentum tensor $T'_{\mu\nu}$ of the scalar field soon after the onset of the phase transition and when ϕ is still zero (that is, the situation that leads to the metric (8.11)), we see from (5.1) (with $T_{\mu\nu} = 0$) and (8.5), we get precisely the Einstein equations with the cosmological constant but zero pressure and density with $\Lambda = 8\pi G V(0)$. Thus the cosmological constant reappears here in quite a different context.

Consider again the situation when the scalar field dominates but it has started deviating from zero. Using (8.8) and (8.9a) we get

$$(\dot{R}/R)^2 \equiv H^2 = (8\pi G/3)[\tfrac{1}{2}\dot{\phi}^2 + V(\phi)]. \tag{8.14}$$

Consider the vanishing of the divergence of the energy–momentum tensor, which gives (2.112) for the Friedmann models, which we write here for convenience:

$$\dot{\varepsilon} + 3(p + \varepsilon)\dot{R}/R = 0.$$

In the present situation of the scalar field we should replace ε, p with ε', p' in this equation in accordance with (8.6), (8.7) and (8.8). Doing this, from (8.8) we get the following equation, after cancelling a factor $\dot{\phi}$:

$$\ddot{\phi} + 3H\dot{\phi} + V' = 0, \quad V' \equiv dV/\partial\phi. \tag{8.15}$$

Equations (8.14) and (8.15) represent the equations which govern the evolution of the scale factor and the scalar field when the latter is the dominant agent of the evolution. In general, exact solutions are difficult to get for any reasonable form of the potential $V(\phi)$. For example, the forms depicted in Figs. 8.2 and 8.3 are given respectively by (8.16a) and (8.16b) below.

$$V(\phi) = \lambda_0\phi^2 + \lambda_1\phi^3 + \lambda_2\phi^4 + V_0, \tag{8.16a}$$

$$V(\phi) = \lambda(\phi^2 - \sigma^2)^2, \tag{8.16b}$$

for suitable values of the constants λ_0, λ_1, λ_2, V_0, λ, σ. However, it is very difficult to find exact solutions of (8.14), (8.15) for the forms (8.16a) and (8.16b) of the potential $V(\phi)$. In the next section we shall consider an exact solution found by the author (Islam, 1990b) for a potential $V(\phi)$ of the sixth degree. For V given by (8.16a) and (8.16b) one usually resorts to an approximation, in one form of which one ignores the $\ddot{\phi}$ term in (8.15), and takes $V(\phi)$ to be given by (8.16b), so that the system is initially at $\phi = 0$ and 'rolls' slowly away from $\phi = 0$, the speed of departure from $\phi = 0$ gradually increasing (Brandenberger, 1987). However, these approximation schemes are unsatisfactory as they sometimes give ambiguous results. For example, Mazenko, Unruh and Wald (1985) argue that in many possible models, conditions for inflation do not obtain in the very early universe. This point of view has been opposed by Albrecht and Brandenberger (1985) who also claim that there are many possible models in which a period of inflation does occur. It would probably be true to say that a completely satisfactory picture for inflation has not yet emerged. See also Pacher, Stein-Schabes and Turner (1987) and Page (1987).

8.4   An exact inflationary solution

In this section we present an exact solution of the coupled scalar field cosmological equations, (8.14) and (8.15), for $V(\phi)$ given as follows:

$$V(\phi) = V_0 + V_1\phi^2 + V_5\phi^5 + V_6\phi^6. \tag{8.17}$$

where the V_i are constants. The solution presented here does not, in fact, satisfy the properties appropriate to the inflation scenario that we have been discussing; for example, $\phi(0) \neq 0$ in this case. Nevertheless, it is of interest because it is an exact solution for a polynomial potential, probably the only exact solution known for such a potential (an exact solution for an exponential potential was found by Barrow (1987)), and the corresponding scale factor does have exponential behaviour over certain ranges of values of t. We will simply state the solution and verify that it is indeed a solution.

Write $q = 3\pi G$, and let n be a constant. Choose the V_i as follows:

$$V_0 = 9n^2/8q; \quad V_1 = n^2; \quad V_5 = -\tfrac{2}{3}n^2q^{3/2}; \quad V_6 = 2n^2q^2/9. \tag{8.18}$$

The solution for ϕ is as follows:

$$\phi(t) = q^{-1/2}\exp(nt)[\exp(nt) + \xi]^{-1} \equiv q^{-1/2}x, \tag{8.19}$$

where ξ is a constant. It is readily verified that

$$\dot{\phi} = nq^{-1/2}(x - x^2); \quad \ddot{\phi} = n^2q^{-1/2}(1 - 2x)(x - x^2). \tag{8.20}$$

With the use of (8.18) and (8.20), Equation (8.14) yields

$$\begin{aligned} H^2 &= \tfrac{8}{9}q(\tfrac{1}{2}\dot{\phi}^2 + V(\phi)) \\ &= n^2(1 + \tfrac{4}{3}x^2 - \tfrac{8}{9}x^3 + \tfrac{4}{9}x^4 - \tfrac{16}{27}x^5 + \tfrac{16}{81}x^6) \\ &= n^2(1 + \tfrac{2}{3}x^2 - \tfrac{4}{9}x^3)^2, \end{aligned} \tag{8.21}$$

so that

$$H = \pm n(1 + \tfrac{2}{3}x^2 - \tfrac{4}{9}x^3). \tag{8.22}$$

If we now substitute for $\ddot{\phi}$, $\dot{\phi}$, H and V' from (8.17), (8.18), (8.20) and (8.22) into (8.15) we find, after some reduction, that the latter is satisfied. The scale factor $R(t)$ can be determined by integrating (8.22), where the negative sign must be taken to satisfy (8.15). The result of the integration is as follows:

$$R(t) = A\exp(-nt)[\exp(nt) + \xi]^{-2/9}\exp\{(2\xi/9)\exp(nt)[\exp(nt) + \xi]^{-2}\},$$

$$\tag{8.23}$$

where A is an arbitrary constant. If we take n to be negative, R has an exponential increase, while $\phi(t)$ goes to zero as t tends to infinity.

The form of the potential (8.17) is that depicted in Fig. 8.5. This can be seen from the fact that the equation $V'(\phi) = 0$ determining the turning points is given as follows:

$$\phi(2V_1 + 5V_5\phi^3 + 6V_6\phi^4) = 0. \tag{8.24}$$

With the use of (8.18), in addition to the root at $\phi = 0$ we get the following equation (in terms of $x = q^{1/2}\phi$):

$$2x^4 - 5x^3 + 3 = 0, \tag{8.25}$$

which has only two real roots, one at $x = 1$, and the other at slightly above $x = 2$. If we assume ζ to be positive, at $t = 0$, x has the value $(1 + \zeta)^{-1}$, and it tends towards zero (for negative n) as t tends to infinity. Thus it goes 'down' the slope from the point B to the point C in Fig. 8.5.

Before closing this section, we will state briefly Barrow's (1987) solution.

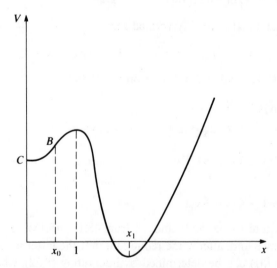

Fig. 8.5. This diagram depicts the form of the potential given by (8.17) and (8.18) in terms of $x = q$. The curve has turning points at $x = 0$, $x = 1$ and $x = x_1$, where x_1 is slightly greater than 2. At $t = 0$, x starts at x_0 and goes to zero as t tends to infinity; $x_0 = (1 + \zeta)^{-1}$. The potential is negative for a finite range of values of the field, but the negative portion does not come into play for the particular solution found here.

This is of the form

$$R(t) = R_0(t/t_0)^b, \tag{8.26a}$$

$$\phi(t) = \phi_0(\log t/\log t_0), \tag{8.26b}$$

$$V(\phi) = V_0 \exp(-\lambda\phi), \tag{8.26c}$$

where t_0, R_0, ϕ_0, V_0, b and λ are suitable constants. Note that this solution gives a power law inflation.

8.5 Further remarks on inflation

In Sections 8.3 and 8.4 we have attempted to give a quantitative description of inflation. It is indicated how at the onset of the phase transition a de-Sitter-like exponential expansion might occur due to the presence of the Higgs field, represented here by a scalar field. However, the further evolution of the scale factor $R(t)$ and the scalar field are not clear due to the difficulty of obtaining exact solutions of the coupled equations, (8.14) and (8.15), for any reasonable potential. In the last section we obtained an exact solution which, though somewhat unrealistic, offers some hope that physically more meaningful solutions might be found.

To discuss phase transitions in the very early universe one must know the so-called 'effective potential' $V(\phi, T)$ as a function of the scalar field ϕ and the temperature T. For temperatures above the critical temperature for the phase transition, the symmetric phase ($\phi = 0$), in which the symmetry among the various interactions is manifest, is the global minimum of the effective potential. One has to derive the effective potential from quantum field theoretic considerations (see, for example, Brandenberger (1985)), but even here one has to resort to approximation schemes. The effective potential used by Albrecht and Steinhardt (1982) and by Linde (1982) is based on the Coleman–Weinberg mechanism (Coleman and Weinberg, 1973) and is given as follows:

$$V(\phi, T) = \alpha\phi^2 - A\phi^4 + B\phi^4 \log(\phi^2/\phi_0^2) + 18(T^4/\pi^2)$$

$$\times \int_0^\infty \mathrm{d}y y^2 \log\{1 - \exp[-(y^2 + 25g^2\phi^2/8T^2)^{1/2}]\}, \quad (8.27)$$

where α, A, B, ϕ_0, g are suitable constants. Figure 8.6 depicts the effective potential given by (8.27) for three typical values of the temperature T.

We will now give some tentative answers to the questions raised at the end of Section 8.2. As regards the monopole problem, the new inflationary model attempts to solve the horizon, magnetic monopole and domain-wall problems in one stroke, namely, by the requirement that before the phase transition the region of space from which the observable universe evolved was much smaller than the horizon distance, so that this region had time to homogenize itself, and because of the inflation from a small portion, the observable universe is expected to have very few monopoles and domain walls, consistent with observation. As regards the matter-antimatter asymmetry, it is possible in some forms of Grand Unified Theories to produce the observed excess of matter over antimatter by elementary particle interactions at temperatures just below the critical temperature of the phase transition, provided certain parameters are suitably chosen. However, there are still many uncertainties in this analysis, but the very possibility of deriving the asymmetry is interesting.

In the form in which the model of the inflationary universe was originally proposed by Guth in 1981 it had a serious defect in that the phase transition itself would have created inhomogeneities to an extent which would be inconsistent with those observed at present. The difficulties with the new inflationary models are, firstly, as already indicated, a completely satisfactory quantitative treatment does not as yet exist and, secondly, in the approximate treatment of the slow-rollover transition, one requires fine tuning of the parameters which seems somewhat

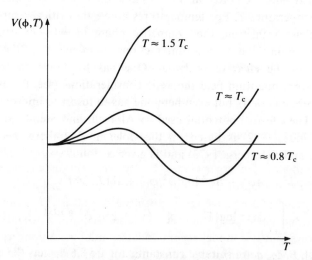

Fig. 8.6. This figure gives the potential represented by (8.27) for three values of the temperature, one slightly above the critical temperature for the phase transition T_c, one near T_c, and a third slightly below T_c. (This is a simplified form of the diagram given in Albrecht and Steinhardt (1982).)

implausible. A great deal of further work needs to be done to clarify the above questions.

Before we end this chapter, we will mention briefly some related points of some importance which we have not been able to deal with in detail. These are firstly the origin of inhomogeneities in the universe that we observe today. One aspect of this problem is similar to the horizon problem. One way to this problem is to consider the inhomogeneities at any time as consisting of perturbations to the smooth background which involve wave-lengths of all scales. However, as one extrapolates this analysis to earlier times, a certain range of the larger wave-lengths becomes longer than the horizon distance, and it becomes a problem as to how these larger wave-lengths arose; in other words, one has to find a mechanism in which wave-lengths larger than the horizon distance are excluded at any time. Another related aspect is the formation of one-dimensional defects during the phase transition; these are known as *cosmic strings* and may have a role to play in the formation or origin of galaxies. We refer the reader to Brandenberger (1987) and Rees (1987) for these questions. Press and Spergel (1989) in particular, explain in a picturesque manner how a field-theoretic description of matter (such as that given by the Lagrangian (8.3)) implies fossilized one-dimensional remnants of an earlier, high temperature phase. Different symmetries of the Langrangians describing possible states of matter in the very early universe give rise to different kinds of remnants, which arise from certain invariant topological properties of space-time. For other consequences of the phase transition in the very early universe, the reader is referred to Miller and Pantano (1989) and to Hodges (1989); the latter author considers domain wall formation. The question of chaotic inflation is considered by Futamase and Maeda (1989) and by Futamase, Rothman and Matzner (1989). Adams, Freese and Widrow (1990) study the problem of the evolution of non-spherical bubbles in the very early universe. The problem of the formation of clusters of galaxies from cosmic strings is investigated by Shellard and Brandenberger (1988). Lastly, we mention 'extended inflation' (La and Steinhardt, 1989; La, Steinhardt and Bertschinger, 1989) in which a special phase transition is not needed, that is, $V(\phi)$ can have a significant barrier between the true and false vacuum phases. Steinhardt (1990) shows that this model accommodates initial conditions leading to $\Omega \leqslant 0.5$. In 'extended inflation' the defects of 'old inflation' are avoided if the effective gravitational constant, G, varies with time during inflation.

9

Quantum cosmology

9.1 Introduction

We saw in the previous chapters that the standard model predicts a singularity sometime in the past history of the universe where the density tends to infinity. In Chapter 6 we also saw there is reason to believe that the existence of singularities may not be a feature peculiar to the highly symmetric Friedmann models, but may exist in any general solution of Einstein's equations representing a cosmological situation. Many physicists think that the existence of singularities in general relativity is unphysical and points to the breakdown of the theory in the extreme situations that singularities purport to represent. Indeed, in these extreme conditions the quantum nature of space-time may come into play, and there have been suggestions that when the quantum theory of gravitation is taken into account, singularities may not arise. However, the quantization of gravitation is notoriously difficult – there does not, at present, exist any satisfactory quantum theory of gravitation, whether the gravitation theory is general relativity or any other reasonable theory of gravity. However, there have been some approximate schemes to try and answer at least partially some of the questions that a quantum theory of gravitation is supposed to answer. One of these schemes is quantum cosmology. We shall only give a brief and incomplete account of quantum cosmology in this chapter, as the technicalities are mostly beyond the scope of this book. This chapter is based mainly on Hartle and Hawking (1983), Hartle (1984, 1986) and Narlikar and Padmanabhan (1983).

We give first a very simple-minded description of quantum theory and see what kind of light an extension of this theory to the cosmological situation may be expected to throw. The quantum theory is supposed to be the basic and fundamental theory which describes all physical phenomena. Classical (non-quantum) physics, including general relativity, is supposed to be an approximation in situations where the action becomes large compared to Planck's constant h, which is of the order of 10^{-27} erg s.

The description we will give here is somewhat crude, but it has the merit of putting in a nutshell the kind of approach the quantum cosmologist has in mind. Consider Fig. 9.1, where (*a*) represents a simple quantum mechanical system, which is described by a single spatial coordinate *x* at any time *t*. The wave function $\psi(x, t)$ represents the quantum mechanical state of the system at time *t*. We will not define here completely what we

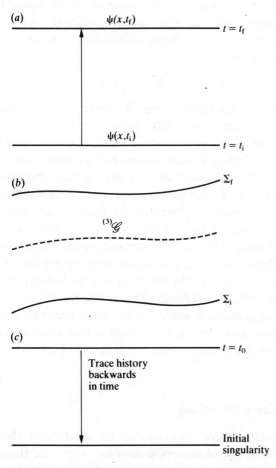

Fig. 9.1. (*a*) In the typical quantum mechanical situation the state of the system is given by the wave function ($\psi(x, t)$) at time $t = t_i$; the Schrödinger equation gives the wave function at a later time $t = t_f$. (*b*) In general relativity the three-geometry is given on a space-like hypersurface Σ_i; the quantum theory of gravitation then gives the probability for the three-geometry on the hypersurface Σ_f. (*c*) If the state of the universe is known at the present time, t_0, the theory should predict the probability of different states that are possible near the 'singularity'.

mean by the 'quantum mechanical state', but suffice it to say that if the wave function is known, all questions of physical interest can be answered with the use of the wave function. It is well known that if the wave function is known at a certain time t_i, the Schrödinger equation then enables us to calculate the wave function at a later time t_f. Extending the simple description given here to the case of a space-time geometry, we might suppose that space-time evolves from a space-like hypersurface Σ_i to another space-like hypersurface Σ_f; the condition of the space-time on any space-like hypersurface is given by the three-geometry on the hypersurface. For example, if we choose coordinates such that the metric may be written as follows

$$ds^2 = dt^2 + g_{jk}\, dx^j\, dx^k, \qquad j, k = 1, 2, 3, \tag{9.1}$$

so that the space-like hypersurfaces may be taken as $t = $ constant, the three-geometry $^{(3)}\mathcal{G}$ at any time is given by the values of g_{jk} at that time. The evolution of the space-time from Σ_i to Σ_f is given classically by Einstein's equations, but in a quantum mechanical description we may ask for the probability of the space-time having any given three-geometry at Σ_f if it has a certain given three-geometry on Σ_i. In practice many difficulties arise; for example, there is the freedom on any surface $t = $ constant (in the metric (9.1)) of carrying out a purely spatial coordinate transformation which does not affect the intrinsic geometry, and one must disentangle the effects of such transformations from the physical evolution of the space-time. The manner in which such a probability amplitude might be found we will consider later. Finally (see (c) of Fig. 9.1), if we did find such a description, we could use the result backwards from the state of the universe at any time t_0 to a period of time near the supposed singularity, to study the quantum mechanical nature of the universe near such a point.

9.2 Hamiltonian formalism

As mentioned earlier, there exists as yet no satisfactory theory for quantizing gravitation. One of the approaches tried so far is the Hamiltonian approach. As is well known, in the Hamiltonian approach one has to single out a particular time coordinate. In general relativity this corresponds to choosing a particular manner of 'slicing' space-time with space-like hypersurfaces. We first give an elementary discussion of the non-relativistic Hamiltonian formalism and the corresponding derivation of the Schrödinger equation; this is mainly to have a simple situation in mind while tackling the more complicated situation later.

We start with a Lagrangian $L(q, \dot{q})$ depending on a generalized coordinate q and its time derivative $\dot{q}$. The equation of motion is found by varying the action S derived from L given as follows:

$$S = \int_{t_1}^{t_2} L(q, \dot{q}) \, dt. \tag{9.2}$$

The condition that the variation $q(t) \to q(t) + \delta q(t)$, with $\delta q(t_1) = \delta q(t_2) = 0$ gives $\delta S = 0$ leads to the Euler–Lagrange equation of motion:

$$\frac{d}{dt} \left(\frac{\partial L}{\partial \dot{q}} \right) - \frac{\partial L}{\partial q} = 0. \tag{9.3}$$

Corresponding to the coordinate q one defines the generalized momentum p as follows:

$$p = \partial L / \partial \dot{q}. \tag{9.4}$$

One then eliminates $\dot{q}$ in favour of p with the use of (9.4), and defines the Hamiltonian as follows:

$$H(p, q) = p\dot{q} - L(q, \dot{q}), \tag{9.5}$$

where it is assumed that $\dot{q}$ has been expressed in terms of p and q. From (9.3), (9.4) and (9.5) it is readily seen that

$$\dot{p} = -\partial H / \partial q, \quad \dot{q} = \partial H / \partial p. \tag{9.6}$$

One defines the Poisson bracket of two functions F, G of p, q as follows:

$$\{F, G\} = \frac{\partial F}{\partial q} \frac{\partial G}{\partial p} - \frac{\partial F}{\partial p} \frac{\partial G}{\partial q}, \tag{9.7}$$

so that (9.6) can be written as follows:

$$\dot{p} = \{p, H\}, \quad \dot{q} = \{q, H\}. \tag{9.8}$$

It is well known that in quantum mechanics the variables q, p, H, etc. become operators in such a manner that the Poisson bracket can be replaced by commutators as follows ($\hbar = h/2\pi$):

$$\{q, H\} \to [q, H] = (i\hbar)^{-1}(qH - Hq), \tag{9.9}$$

etc., so that, by comparing with the equation of motion of a free particle derived from the Lagrangian $L = \frac{1}{2}m\dot{q}^2$ (given by $\dot{q} = p/m$), we have the following commutation relation between q and p in quantum mechanics:

$$qp - pq = i\hbar. \tag{9.10}$$

F

This implies that p can be expressed as the operator $p = -i\hbar\,\partial/\partial q$. One can also show that the energy E can be replaced by the operator $i\hbar\,\partial/\partial t$. It is readily seen that if the Lagrangian is given by

$$L = \tfrac{1}{2}m\dot{q}^2 - V(q), \tag{9.11}$$

which represents a particle moving in a potential $V(q)$, the Hamiltonian is given by

$$H = (2m)^{-1}p^2 + V(q), \tag{9.12}$$

so that, since the Hamiltonian represents the energy, we get the Schrödinger equation by applying both sides of (9.12) as operators to the wave function $\psi(q, t)$, whose modulus square $|\psi(q, t)|^2$ represents the probability density of finding the particle at q at time t; in fact $|\psi|^2\,dq$ is the probability of the particle being between q and $q + dq$:

$$i\hbar\frac{\partial\psi}{\partial t} = \left[-\frac{\hbar^2}{2m}\frac{\partial^2}{\partial q^2} + V(q)\right]\psi. \tag{9.13}$$

Consider now the Lagrangian for several particles given by

$$L = \sum_r L(q_r, \dot{q}_r), \tag{9.14}$$

where q_r represents the coordinate of the rth particle, and the generalized canonical momentum corresponding to q_r is given by

$$p_r = \partial L/\partial\dot{q}_r. \tag{9.15}$$

The corresponding Hamiltonian is given by

$$H(p_r, q_r) = \sum_r p_r\dot{q}_r - \sum_r L(q_r, \dot{q}_r). \tag{9.16}$$

We now consider the case of the Lagrangian of a field – this is like replacing the coordinate of the rth particle $q_r(t)$ by $\phi(\mathbf{x}, t)$, so that the index r is replaced by the spatial position $\mathbf{x}$ in a suitably limiting sense. The Lagrangian in this case is a function of the fields $\phi(\mathbf{x}, t)$ and the time and spatial derivatives $\partial_\mu\phi(\mathbf{x}, t)$. The sum over particles becomes an integral over the spatial coordinates ($\mathscr{L}$ is the Lagrangian density):

$$L(t) = \int \mathscr{L}(\phi, \partial_\mu\phi)\,d^3x. \tag{9.17}$$

One defines a field $\pi(\mathbf{x}, t)$ canonically conjugate to $\phi(\mathbf{x}, t)$ as follows:

$$\pi = \partial\mathscr{L}/\partial\dot{\phi}. \tag{9.18}$$

In analogy with the particle case one defines the Hamiltonian:

$$H(t) = \int (\pi\dot\phi - \mathcal{L}) \, d^3x. \tag{9.19}$$

One can also make a simple-minded extension of the idea of a wave function to that of a wave functional $\Psi[\phi, t]$ which is a functional of the fields and a function of the time. It can be interpreted as saying that $|\psi[\phi, t]|^2 \delta\phi$ is the probability of finding the field configuration between ϕ and $\phi + \delta\phi$ at time t. In analogy with the particle case, the Schrödinger equation in this case is given by

$$H\Psi[\phi, t] = i\hbar \, \partial\Psi/\partial t. \tag{9.20}$$

Here H is given by (9.19), where the $\dot\phi$ has to be eliminated in favour of π using (9.18), and later π has to be replaced by the operator $-i\hbar \, \delta/\delta\phi$, that is, $-i\hbar$ times the functional derivative with respect to ϕ.

A functional is a number which depends on a function on the whole domain of its definition. Restricting to one variable x, a functional F of a function $A(x)$ may be given by

$$F[A] = \int f(x)A(x) \, dx, \tag{9.21}$$

where $f(x)$ is a fixed function. The fundamental relation for taking functional derivatives is the following one

$$\delta A(x)/\delta A(x') = \delta(x - x'), \tag{9.22}$$

where on the right we have the Dirac delta function. With the use of (9.22) we find readily that

$$\frac{\delta F}{\delta A(x')} = \int f(x) \frac{\delta A(x)}{\delta A(x')} \, dx = \int f(x)\delta(x - x') \, dx = f(x') \tag{9.23}$$

In a similar manner with the use of (9.22) and the rules of ordinary differentiation, one can evaluate all functional derivatives.

9.3   The Wheeler–De Witt equation

In quantum gravity one can derive an equation similar to the Schrödinger functional equation (9.20) which is known as the Wheeler–De Witt equation. This equation is best derived from the path integral formalism, which we will consider in the next section. We shall not give the derivation here but only write down the equation itself and give a brief description of it.

As mentioned earlier, there are many subtleties which we will not consider. One of these is the manner in which space-time should be sliced to give a series of appropriate three-geometries, in which there remains the problem of dealing with the freedom of carrying out spatial transformations. One of the conditions that go into the derivation of the Wheeler–De Witt equation (that given, for example, by Hartle and Hawking (1983)), is that the universe should be closed, so that the space-like sections are compact. We use units with $\hbar = c = 1$ and introduce coordinates so that the space-like hypersurfaces are $t = $ constant and the metric is written as follows:

$$ds^2 = (N^2 - N_i N^i)\, dt^2 - 2N_i\, dx^i\, dt - h_{ij}\, dx^i\, dx^j. \quad i, j = 1, 2, 3. \quad (9.24)$$

N, N^i are functions of space-time with $N_i = h_{ij} N^j$. K_{ij} is the extrinsic curvature of the three-surface $t = $ constant given as follows:

$$K_{ij} = -n_{i;\,j}, \quad (9.25)$$

where n^i is the spatial part of the unit normal to the hypersurface, $t = $ constant and the semi-colon denotes covariant derivative as in (2.6b). Equation (9.25) can be written as follows:

$$K_{ij} = N^{-1}(-\tfrac{1}{2}\dot{h}_{ij} + \nabla_j N_i), \quad (9.26)$$

where ∇_j denotes covariant derivative with respect to the three-metric h_{ij}. The momentum canonically conjugate to h_{ij} is given as follows in terms of K_{ij} and its trace $K = K_i^{\,i}$:

$$\pi_{ij} = -h^{1/2}(K_{ij} - h_{ij}K), \quad (9.27)$$

where h is the determinant of the metric h_{ij}. The wave functional in this case is a functional $\Psi[h_{ij}]$ of the three-metric h_{ij} and is related to the probability of finding the space-like hypersurface with the given three-metric. One can find an expression which is equivalent to the Hamiltonian, and if one replaces the π^{ij} with the operator $-i\delta/\delta h_{ij}$ in it, one gets the Wheeler–De Witt equation

$$\left(-G_{ijkm}\frac{\delta^2}{\delta h_{ij}\delta k_{km}} - {}^3R h^{1/2} + 2\Lambda h^{1/2}\right)\Psi[h_{ij}] = 0. \quad (9.28)$$

Here

$$G_{ijkm} = \tfrac{1}{2}h^{-1/2}(h_{ik}h_{jm} + h_{im}h_{jk} - h_{ij}h_{km}), \quad (9.29)$$

^3R is the scalar curvature for the three-metric, and Λ the cosmological constant. Equation (9.28) corresponds to the stationary form of Schrödinger's equation given by $H\Psi = E\Psi$. The tensor G_{ijkm} is, in fact, the metric in the 'superspace', which is the space of all three-geometries. In (9.28) we have also ignored the matter fields, for which one would get additional terms. The freedom to carry out spatial transformations of the three-metric gives additional constraints which the wave functional must satisfy – these are familiar in gauge theories in the so-called Gauss relations. This completes our brief discussion of the Wheeler–De Witt equation. We will now consider the equivalent path integral approach, for which we first give a brief account of path integrals.

9.4   Path integrals

In recent years path integrals have been used increasingly in the formulation of gauge theories and other aspects of physics. The originator of the method of path integrals was Feynman (1948) (see also Feynman and Hibbs (1965)). There are many introductory accounts of path integrals (see, for example, Taylor (1976)). We will give the bare essentials here (see Narlikar and Padmanabhan (1983)).

A convenient way of introducing path integrals is to compare the formulation of the equations of motion of a free particle in classical mechanics and quantum mechanics as expressed in terms of path integrals. If the position vector of the particle is $\mathbf{r} = (x, y, z)$, its equation of motion is given by

$$m\ddot{\mathbf{r}} = 0, \tag{9.30}$$

Suppose the particle is at $\mathbf{r}_i$ at time t_i (the initial time) and at $\mathbf{r}_f$ at time t_f (the final time). It is easy to see that in the intervening period $t_i < t < t_f$ the position vector is given by

$$\mathbf{r}(t) = \mathbf{r}_i + \left(\frac{t - t_i}{t_f - t_i}\right)(\mathbf{r}_f - \mathbf{r}_i) \equiv \bar{\mathbf{r}}(t). \tag{9.31}$$

Quantum mechanically, if the particle is at $\mathbf{r}_i$ at time t_i, one can only give a *probability amplitude* for finding the particle at $\mathbf{r}_f$ at time t_f; this is given as follows:

$$K(\mathbf{r}_f, t_f; \mathbf{r}_i, t_i) = [m/2\pi i\hbar(t_f - t_i)]^{3/2} \exp[im(\mathbf{r}_f - \mathbf{r}_i)^2/2\hbar(t_f - t_i)]. \tag{9.32}$$

The connection between (9.31) and (9.32) is established by saying that classically the particle follows the definite path $\bar{\mathbf{r}}(t)$ given by (9.31) whereas quantum mechanically the particle can take any path that is allowed by causality; there is a certain amplitude associated with each path $\mathbf{r}(t)$ and to get the complete amplitude to find the particle at $\mathbf{r}_f$ at time t_f one has to sum over all paths weighted by the amplitude for the path (see Fig. 9.2). The amplitude associated with the path $\mathbf{r}(t)$ is given by $\exp[(i/\hbar)S]$ where S is the classical action associated with the path, given by (9.2) (with $\mathbf{r}(t)$ instead of q; that is, instead of the one coordinate q we have three coordinates $\mathbf{r}$). The sum is then taken over all paths; this sum is a kind of integral over all functions $\mathbf{r}(t)$, which can be defined by a limiting procedure in which the time interval $[t_i, t_f]$ is divided into n equal parts so that the weight function becomes a function of the variables $\mathbf{r}(t_m)$, where t_m is a typical instant at which the division of the interval $[t_i, t_f]$ occurs. One then integrates over the $\mathbf{r}_m \equiv \mathbf{r}(t_m)$, and takes the limit as n tends to infinity to get the complete amplitude to arrive at $\mathbf{r}_f$ at time t_f (see, for example, Feynman and Hibbs (1965) for details). For example, if one starts with the classical action which gives rise to (9.30), namely,

$$S = \int_{t_i}^{t_f} \tfrac{1}{2}m\dot{\mathbf{r}}^2 \, dt, \qquad (9.33)$$

one arrives at the amplitude (9.32) by adopting this limiting procedure. Symbolically, this integral can be written as follows:

$$K(\mathbf{r}_f, t_f; \mathbf{r}_i, t_i) = \int \exp\{(i/\hbar)S[\mathbf{r}(t)]\}\mathscr{D}\mathbf{r}(t). \qquad (9.34)$$

Note that here S is not a function but a functional of $\mathbf{r}(t)$; for this reason this integral is also called a functional integral. Here the symbol $\mathscr{D}\mathbf{r}(t)$ means the integral is a sum over all functions $\mathbf{r}(t)$ in the sense explained in Feynman and Hibbs (1965).

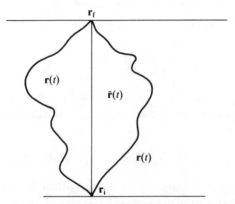

Fig. 9.2. Classically the particle follows the path $\bar{\mathbf{r}}(t)$; quantum mechanically the particle can follow any of the paths $\mathbf{r}(t)$, but each path is weighted by the amplitude $\exp(i/\hbar)S$, where S is the classical action associated with the path $\mathbf{r}(t)$.

By dividing every path at a certain instant t, one can derive from (9.34) the following relation:

$$K(\mathbf{r}_f, t_f; \mathbf{r}_i, t_i) = \int K(\mathbf{r}_f, t_f; \mathbf{r}, t) K(\mathbf{r}, t; \mathbf{r}_i, t_i) \, d^3 r, \qquad (9.35)$$

for any time $t_i \leqslant t \leqslant t_f$. Similarly, one can show that if the state of the particle at time t_i is represented by the wave function $\psi_i(\mathbf{r}_i, t_i)$, then its final wave function $\psi_f(\mathbf{r}_f, t_f)$ is given as follows:

$$\psi(\mathbf{r}_f, t_f) = \int K(\mathbf{r}_f, t_f; \mathbf{r}_i, t_i) \psi_i(\mathbf{r}_i, t_i) \, d^3 r_i. \qquad (9.36)$$

This relation can be verified explicitly for the free particle wave function $\psi(\mathbf{r}_i, t_i) \sim \exp[(i/\hbar)(Et_i - \mathbf{p} \cdot \mathbf{r}_i)]$, with a similar expression for $\psi_f(\mathbf{r}_f, t_f)$ where $E = p^2/2m$ if one uses the K given by (9.32). In fact this confirms that (9.32) is the correct free particle amplitude.

By analogy with the above case, one can consider the case of a space-time geometry, in which Σ_i, Σ_f represent respectively an initial and a final space-like hypersurface (see Fig. 9.1(b)), and one asks for the probability amplitude for a certain three-geometry $^{(3)}\mathcal{G}_f$ on Σ_f given the three-geometry $^{(3)}\mathcal{G}_i$ on Σ_i. In this case the classical 'path' is the solution given by Einstein's equations, but the contributions to the probability amplitude come from all four-geometries which are not necessarily solutions of Einstein's equations. Symbolically this can be represented by

$$K\{^{(3)}\mathcal{G}_f, \Sigma_f; {}^{(3)}\mathcal{G}_i, \Sigma_i\} = \int \exp\{(i/\hbar) S[g]\} \mathcal{D}g, \qquad (9.37)$$

in analogy with (9.34). Here $S[g]$ is the action for gravitation (see, e.g., (9.38) below) and the functional integration is over all four-geometries connecting Σ_i and Σ_f. There are, of course, many complexities hidden in (9.37). For example, one has to take into account that some four-geometries are simply transforms of each other. Presumably these can be taken into account by a generalization of the method of Faddeev and Popov (1967) which is used in Yang–Mills type gauge theories and which effectively amounts to dividing out an infinite gauge volume. Secondly, the actual evaluation of the path integral (9.37) for any given situation presents prohibitive problems. Nevertheless, the conceptual simplicity of (9.37) is striking. In the next section we will examine how some information can be extracted from (9.37) with some simplifying assumptions.

We end this section with some remarks about the classical limit of the path integral (9.34). Classical physics is valid when the action of the classical system is large compared to $\hbar$; note that S and $\hbar$ have the same dimensions so that $S/\hbar$ is a pure number. Thus for a classical system the phase of the exponential in (9.34) is large for most paths, so that a small variation in the path causes a relatively large variation in the phase with the result that, because of the oscillation of the exponential, the contributions from neighbouring paths cancel each other. The only paths which contribute substantially in this case are those for which the action does not vary much with the variation in the paths. These are given by paths near the one which gives $\delta S = 0$, which, of course, yields the classical path $\bar{\mathbf{r}}(t)$. Thus in the limit of vanishing $\hbar$ we get the classical path. The interesting thing is that the same argument applied to (9.37) yields the Einstein equations in the classical limit, these equations being given by $\delta S_{\mathrm{g}} = 0$, where S_{g} is given as follows, inserting c:

$$S_{\mathrm{g}} = (c^3/16\pi G) \int_{\mathscr{V}} R(-g)^{1/2}\, \mathrm{d}^4 x, \qquad (9.38)$$

where $\mathscr{V}$ is the space-time region under consideration. The scalar curvature R of the space-time has dimensions (length)$^{-2}$. Thus if we take $R \sim L^{-2}$, where L is the characteristic length, and the four-volume $\mathscr{V}$ is of dimension L^4, we find

$$S_{\mathrm{g}} = c^3 L^2/16\pi G. \qquad (9.39)$$

Thus the action S_{g} becomes comparable to $\hbar$ if the linear size of the universe is (ignoring the numerical factor 16π)

$$L_{\mathrm{P}} = (G\hbar/c^3) \sim 1.6 \times 10^{-33}\ \mathrm{cm}, \qquad (9.40)$$

which is the so-called *Planck length*.

We note finally that the Schrödinger equation can be derived from (9.36) by making $t_{\mathrm{f}} - t_{\mathrm{i}}$ infinitesimally small.

9.5   Conformal fluctuations

We have seen that in the path integral (9.37), the sum involves space-times which do not necessarily satisfy Einstein's equations. In practice to include all such space-times is a formidable task. One simplification that has been tried is to consider only geometries which are conformal to the classical solutions, that is, solutions of Einstein's equations. Suppose that for a given action (9.38) we have a classical solution given by the metric

$$\mathrm{d}\bar{s}^2 = \bar{g}_{ik}\, \mathrm{d}x^i\, \mathrm{d}x^k, \qquad (9.41)$$

for the region which lies between the space-like hypersurfaces Σ_{i} and Σ_{f}

(see Fig. 9.1(*b*)). Non-classical paths also contribute to (9.37), but we consider only those paths which are conformally related to (9.41), that is, only those metrics which are of the following form:

$$ds^2 = \Omega^2 \, d\bar{s}^2 = \Omega^2 \bar{g}_{ik} \, dx^i \, dx^k, \tag{9.42}$$

where Ω is an arbitrary function of space-time. Since Einstein's equations are not conformally invariant, except in the trivial case $\Omega = $ constant, (9.42) represents a non-classical path between Σ_i and Σ_f. There are other ways of generating non-classical paths, but the merit of (9.42) is that null geodesics are conformally invariant, so the light cone structure of space-time is preserved by such paths. We write

$$\phi = \Omega - 1, \tag{9.43}$$

so that ϕ represents the conformal fluctuation around the classical path. We shall only give the results of the consideration of conformal paths, and refer to Narlikar and Padmanabhan (1983) for the details. We take the classical geometry to be that of Friedmann cosmologies, which we write as follows

$$d\bar{s}^2 = dt^2 - Q^2(t)[dr^2/(1 - kr^2) + r^2(d\theta^2 + \sin^2\theta \, d\phi^2)]. \tag{9.44}$$

We consider the state of the universe at the initial epoch t_i to be given by a wave packet with spread Δ_i, as follows:

$$\psi_i(\phi, t_i) = (2\pi\Delta_i^2)^{-1/4} \exp(-\phi^2/4\Delta_i^2). \tag{9.45}$$

It is shown by Narlikar (1979) and by Narlikar and Padmanabhan (1983) that if conformal paths are taken into account, the wave packet (9.45) evolves to the one given by a similar expression to (9.45) except that Δ_i is replaced by Δ_f given as follows (see (9.36) and Fig. 9.3):

$$\Delta_f = (2\pi T/3VQ_iQ_f)[1 + (3V/2\pi T)\Delta_i^2 Q_i^2(1 + TQ_iH_i)^2]^{1/2}, \tag{9.46}$$

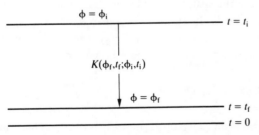

Fig. 9.3. For the conformal fluctuation of Friedmann cosmologies we reverse the time and take the final time t_f to be earlier than the initial time t_i, with the former near the singularity at $t = 0$.

where V is the coordinate volume of the region under consideration, given by $r \leqslant r_b$, where r is the radial coordinate occurring in (9.44) and T, H^i are defined as follows:

$$T = \int_{t_f}^{t_i} du/Q(u), \quad H_i = \dot{Q}(t_i)/Q(t_i). \tag{9.47}$$

The important thing to notice is that as t_f tends to zero, that is, as we approach the singularity, Δ_f goes as Q_f^{-1}, and so diverges. Thus it appears that in the limit of the classical singularity quantum conformal fluctuations diverge. Thus the classical solution, which can be regarded as the 'average' of the wave packet, is no longer reliable near the singularity. Narlikar and Padmanabhan (1983) further find indications that quantum conformal fluctuations may prevent a space-time singularity and also may eliminate the appearance of a particle horizon.

There are many approximations involved in the above considerations and so, many uncertainties. It is not clear to what extent the claims made in the above paragraph are valid. The important thing to notice here, however, is that the formalism of this chapter seems to provide a handle with which these interesting questions can be meaningfully tackled. There is obviously a long way to go before definitive answers can be given to such questions. The above work has been generalized by Joshi and Narlikar (1986) to cases where the state of the universe is defined by wave functionals that are not necessarily wave packets, with similar results.

To end this section we consider as an illustration conformal perturbation of flat space-time. For this we first consider transform of S_g given by (9.38) under the conformal transformation (9.42). Putting $c = \hbar = 1$, S_g is transformed to S_g' given by the following expression:

$$S_g' = (16\pi G)^{-1} \int_{\mathscr{V}} (\Omega^2 \bar{R} - 6\Omega_\mu \Omega^\mu)(-\bar{g})^{1/2} \, d^4 x, \quad \Omega_\mu = \frac{\partial}{\partial x^\mu} \Omega, \tag{9.48}$$

where $\bar{R}$ is the scalar curvature derived from the metric $\bar{g}_{ij}$, and $\bar{g}$ is the determinant of this metric. If we now specialize the metric $\bar{g}_{ik}$ to that of flat space given by $\eta_{\mu\nu}$ with $-\eta_{00} = \eta_{11} = \eta_{22} = \eta_{33} = -1$, with $\eta_{\mu\nu} = 0$ when $\mu \neq \nu$, the action S_g' reduces to the following one:

$$S_\Omega = -(3/8\pi G) \int \Omega_\mu \Omega^\mu \, d^4 x. \tag{9.49}$$

We apply the formalism developed in (9.17)–(9.20). The Lagrangian density for (9.49) is given as follows:

$$\mathscr{L} = (-3/8\pi G)\Omega_\mu \Omega^\mu = (-3/8\pi G)(\dot{\Omega}^2 - (\nabla\Omega)^2), \tag{9.50}$$

so that the momentum density canonically conjugate to Ω is given by

$$\partial\mathscr{L}/\partial\dot{\Omega} = 2k'\dot{\Omega} = \pi, \quad k' \equiv (-3/8\pi G). \tag{9.51}$$

The Hamiltonian density is given as follows:

$$\mathscr{H} = \pi\dot{\Omega} - \mathscr{L} = k'[\dot{\Omega}^2 + (\nabla\Omega)^2] = (4k')^{-1}\pi^2 + k'(\nabla\Omega)^2. \tag{9.52}$$

The corresponding Schrödinger equation is (replacing π by $-i\delta/\delta\Omega$):

$$(4k')^{-1}\int[-(\delta/\delta\Omega)^2 + 4k'^2(\nabla\Omega)^2]\,\mathrm{d}^3x\Psi[\Omega] = E\Psi[\Omega]. \tag{9.53}$$

This equation is similar to the one that obtains in quantum electrodynamics of the pure electromagnetic field, for which the solution is well known (see, for example, Rossi and Testa (1984), Hartle (1984), Islam (1989); also Feynman and Hibbs (1965)). The ground state solution of (9.53) can be written as follows:

$$\Psi[\Omega] = N\exp\left[\left(-\frac{3}{8\pi^3 L_{\mathrm{P}}^2}\right)\int\int\frac{\nabla_x\Omega\cdot\nabla_y\Omega}{(\mathbf{x}-\mathbf{y})^2}\,\mathrm{d}^3x\,\mathrm{d}^3y\right], \tag{9.54}$$

which gives the probability amplitude for detecting a conformal factor in flat space. This expression implies that large deviations from flat space can occur at Planck length scales L_{P}. This is usually referred to as the 'foam' structure of space-time.

9.6 Further remarks about quantum cosmology

We end this chapter by mentioning some further developments. One significant one is the proposal for the wave function of the 'ground state' of the universe, put forward by Hartle and Hawking (1983), which we describe here briefly. An interesting aspect of any quantum mechanical theory is the ground state or the state of minimum excitation. In terms of path integrals, the ground state at $t = 0$ can be defined by

$$\psi_0(x, 0) = N\int\exp\{-I[x(\tau)]\}\mathscr{D}x(\tau), \tag{9.55}$$

where the time integral in the action S has been transformed by $t \to -i\tau$, to make the path integral well defined (this does not, in general, affect its value) and iS has been replaced by $-I$. The function $x(\tau)$ represents all paths which end at $x(0) = x$ at $t = \tau = 0$. (A proof of (9.55) can be found in Hartle and Hawking (1983).)

In the case of closed universes, which Hartle and Hawking consider, it is not appropriate to define the ground state as the state of lowest energy, as there exists no natural definition of energy for a closed universe. In fact, the total energy of a closed universe may be zero – the gravitation and matter energies cancelling each other. It might be reasonable, however, to define a state of minimum excitation corresponding classically to a geometry of high symmetry. In analogy with (9.55), Hartle and Hawking propose the following expression as the ground state wave function of a closed universe:

$$\psi_0[h_{ij}] = N \int \exp(-I_{\mathrm{E}}[g]) \mathcal{D}g, \tag{9.56}$$

where I_{E} is the Euclidean action for gravity (obtained by carrying out the transformation $t \to -i\tau$ in S_{g} given by (9.38)) and including the cosmological constant Λ. They are able to work out the path integral using certain simplifying assumptions, and find that the ground state corresponds to de Sitter space in the classical limit. They also find excited states which yield universes which start from zero volume, reach a maximum and collapse, but which also have a non-zero (but small) probability of tunnelling through a potential barrier to a de Sitter type of continued expansion.

We have glossed over several complexities earlier in the chapter. One of these is the problem of 'operator ordering' in (9.28) where a simple ordering between π_{ij} and h_{ij} has been used. Another possibility for the first term in (9.28) would have been, for example,

$$(2h^{1/2})^{-1}(\delta/\delta h_{ij})h^{1/2}G_{ijkm}(\delta/\delta h_{km}). \tag{9.57}$$

A term q^2p, for example, in the classical Hamiltonian, can become q^2p, qpq, or pq^2 and one has to use other other considerations to decide which is the correct one, as quantum mechanically these are apparently distinct possibilities, since here q, p are non-commuting.

Coleman and Banks (see Schwarzschild (1989), and references therein) have considered a modification of the Hartle–Hawking formalism given by (9.55) and (9.56) in which the path integrals are not only over the entire history of the present universe, but also over the full manifold of all universes connected by wormholes (see, for example, Misner, Thorne and Wheeler (1973), p. 1200). In the resulting analysis they find an explanation of the vanishing of the cosmological constant (see also Weiss (1989)).

We have attempted to provide here the bare minimum of the subject of quantum cosmology. It is hoped that this will enable the reader to follow the more specialized material in the papers cited here (see particularly Hartle (1986) and the papers cited there).

10

The distant future of the universe

10.1   Introduction

In the previous chapters we have considered in some detail the 'standard' model of the universe. It is pertinent to ask what the prediction of the standard model is for the distant future of the universe. The future of the universe has been the subject of much speculation, in one form or another, from time immemorial. It is only in the last few decades that enough progress has been achieved in cosmology to study this question scientifically. In this chapter we shall attempt to provide an account of – or at any rate limit the possibilities for – the distant future of the universe, on the basis of the present state of knowledge. We refer the reader to Rees (1969), Davies (1973), Islam (1977, 1979a,b, 1983a,b), Barrow and Tipler (1978) and Dyson (1979) for more material on this topic. This chapter is based mostly on the papers by Islam and Dyson.

The distant future of the universe is dramatically different depending on whether it expands forever, or it stops expanding at some future time and recollapses. In the earlier chapters we have considered in detail the conditions under which these possibilities are likely to arise. As galaxies are the basic constituents of the universe, to examine the distant future of the universe we must consider the long term evolution of a typical galaxy. We will first assume that we are in an open universe, or at any rate, that an indefinite time in the future is available. It is worth noting that by taking the mass density of the universe to be above but sufficiently close to the critical density, we can get models of the universe which have a finite but arbitrarily long life-time.

10.2   Three ways for a star to die

In any amount of matter there is a tendency for the matter to collapse towards the centre of mass due to the gravitational attraction of different parts for each other. In a star this inward force is balanced by the release of energy during nuclear burning in which hydrogen is converted into helium and helium into heavier nuclei. At this stage the material of the star can be approximated by an ideal gas, in which the pressure p is related to its temperature T and number density n by the relation:

$$p = nkT, \tag{10.1}$$

where k is Boltzmann's constant (there should not be any confusion with the k used in the Robertson–Walker metric). As the star loses energy and its temperature decreases, this thermal energy, after a few billion years, is insufficient to balance the inward force of gravity. The star contracts and becomes more dense so that the electrons are eventually stripped off the atoms and run about freely in the material of the star. They then exert a Fermi pressure due to the Pauli exclusion principle. When the density is about $5 \times 10^6 \, \mathrm{g\,cm^{-3}}$ this electron degeneracy pressure is given by, restoring c (Chandrasekhar, 1939)

$$p \sim hcn^{4/3}. \tag{10.2}$$

At lower densities p is proportional to $n^{5/3}$.

For a spherically symmetric star, p and the mass density ρ satisfy the equation of hydrostatic equilibrium at radius r .

$$dp/dr = -G[m(r)/r^2]\rho, \tag{10.3}$$

where $m(r)$ is the mass inside radius r. One can show that in order to support itself against collapse the pressure p_c at the centre must be

$$p_c \sim GM^{2/3}\rho^{4/3}, \tag{10.4}$$

where M is the total mass of the star. Thus the pressure available at high densities (10.2) and the pressure needed for support have the same dependence on n (since ρ is proportional to n). It can be shown that when M is less than about 1.4 times the mass of the Sun, the electron degeneracy pressure can permanently halt collapse (Chandrasekhar, 1935, 1939) and one gets white dwarfs whose size is roughly that of the Earth. These eventually become cold and stop radiating altogether to become what are sometimes called 'black dwarfs'. The nuclei in these stars are mostly those of iron, since the latter has the most stable nucleus.

When the mass of the star is greater than 1.4 solar masses, or if there is a sudden inward pressure due to an explosion of the outer layers, the electron degeneracy pressure is insufficient to balance gravity. The star continues to collapse and becomes more dense until the electrons are squeezed into the protons of the nuclei to become neutrons and different nuclei coalesce until the star becomes a giant nucleus – a neutron star. If the mass of the star is less than a certain critical mass M_c (this is about 2–3 solar masses) the neutron degeneracy pressure and the forces of nuclear interactions are sufficient to balance gravity. To find M_c one must appeal to general relativity, since Newtonian theory is inadequate for the strong fields generated by the neutron stars. For the latter (10.3) is replaced by

$$\mathrm{d}p/\mathrm{d}r = G(\rho + p/c^2)[m(r) + 4\pi r^3 p/c^2]/\{r[r - 2Gm(r)/c^2]\}. \qquad (10.5)$$

Equation (10.5) implies that more pressure is needed to support a star for strong fields than is implied by Newtonian theory. Neutron stars are the pulsars, discovered in 1967, of which more than three hundred have been found since the original discovery (Hewish *et al*, 1968).

When the mass of the star is greater than M_c after shedding any mass, even neutron degeneracy pressure and the forces of nuclear interactions are insufficient to halt the collapse. In this case there is no known force which can halt the collapse and it is assumed that the star continues to collapse until it gets literally to a point – into a space-time singularity akin to the space-time singularity of the very early universe, about the nature of which, as seen earlier in this book, there is a great deal of uncertainty. This collapse results in a black hole which is a spherical region of radius $2GM/c^2$, where M is the mass of the star. If M is ten times the solar mass then this radius – the Schwarzschild radius – is about 20–30 km. The surface of the sphere of the Schwarzschild radius is called the horizon and the spherical region is called a black hole because once the star collapses to within this region nothing – not even light – can escape. There may, of course, be radiation from infalling matter just before the matter enters the region. The black hole may be detected by such radiation and also by its gravitational influence on nearby stars, etc. (see, for example, Thorne (1974)).

The above three final states, namely, those of black dwarf, neutron star and black hole occur for masses not too small compared to the mass of the Sun. For smaller bodies such as the Earth and the Moon or a piece of rock, gravity can be balanced indefinitely by the ordinary pressure that matter exerts in resisting being compressed.

10.3   Galactic and supergalactic black holes

Consider the fate of a typical galaxy assuming we have an indefinite period ahead. All stars will ultimately be reduced to black dwarfs, neutron stars or black holes. As the galaxy will be losing energy by radiation all the time, including the thermal energy of any hot interstellar gas, given sufficient time the galaxy will eventually consist of a gravitationally bound system of black holes, neutron stars, black dwarfs and cold interstellar matter in the form of planets, asteroids, meteorites, dust, etc. From the average energy and luminosity of a typical galaxy one can deduce that the time scale to arrive at this state will be anything between 10^{11} and 10^{14} years.

This situation will continue for thousands of billions of years without any significant changes within galaxies, but galaxies which are not in the same cluster will continue to recede from each other. The next significant change in a galaxy will take much longer than earlier changes such as stars becoming black holes, etc. The stars (henceforth by 'stars' we mean black dwarfs, neutron stars or stellar size black holes) in the galaxy will eventually tend to form a dense central core with an envelope of low density. The long term evolution of such a system is very difficult to predict accurately (see, for example, Saslaw (1973) and Saslaw, Valtonen and Aarseth (1974)). Some stars, if they are involved in close three-body or many-body encounters, may be thrown out of the galaxy altogether. Such encounters are relatively rare in time scales of a few billion years. The time scales over which such processes dominate can be worked out as follows (Dyson, 1979). If a galaxy consists of N stars of mass M in a volume of radius R, their root mean square velocity will be of the order

$$v = (GNM/R)^{1/2}. \tag{10.6}$$

The cross-section for close collision is

$$\sigma = (GM/v^2)^2 = (R/N)^2, \tag{10.7}$$

and the average time spent by a star between two collisions is

$$t_{av} = (\rho v \sigma)^{-1} = (NR^3/GM)^{1/2}, \tag{10.8}$$

where ρ is the density of stars in space. For a typical galaxy $N = 10^{11}$, $R = 3 \times 10^{17}$ km, so

$$t_{av} = 10^{19} \text{ years.} \tag{10.9}$$

Dynamical relaxation of the galaxy takes about 10^{18} years. The combined effect of close collisions and dynamical relaxation is to produce a dense central core which eventually collapses to a single black hole, while stars from the outer regions evaporate in a time scale of that given by (10.9). The number of stars that will escape is very difficult to determine; perhaps 99%. Thus in about 10^{20} years or somewhat longer the original galaxy will be reduced to a single 'galactic' black hole of about 10^9 solar masses, while stray stars and other small pieces of matter thrown out of the galaxy will be wandering singly in the intergalactic space.

It is likely that a cluster of galaxies will continue to be gravitationally bound as the expansion of the universe proceeds. Through long term dynamical evolution as described above the cluster will also eventually reduce to a single 'supergalactic' black hole of about 10^{11} or 10^{12} solar masses, a large fraction of the stars having evaporated.

This process of the transformation of the original galaxy into a single black hole may be slightly affected by gravitational radiation. When a number of stars go round each other, they radiate gravitational waves, thus lose energy and become more tightly bound. The time scale over which this process has a significant effect on the galaxy is anything from 10^{24} to 10^{30} years (Islam, 1977; Dyson, 1979). Thus the effects of dynamical evolution will be more dominant than those of gravitational radiation.

10.4 Black-hole evaporation

According to the laws of classical mechanics, a black hole will last forever. It was shown by Hawking (1975) that when quantum phenomena are taken into account, a black hole is not perfectly black but gives off radiation such as electromagnetic waves and neutrinos. 'Empty' space is actually full of 'virtual' particles and antiparticles that come into existence simultaneously at a point in space, travel a short distance and come together again, annihilating each other. The energy for their existence can be accounted for by the uncertainty principle. In the neighbourhood of the horizon of a black hole it might happen that one particle from a virtual pair falls into the black hole with negative energy, while its partner, unable to annihilate, escapes to infinity with positive energy. The negative energy of the infalling particle causes a decrease in the mass of the black hole. In this manner the black hole gradually loses mass and becomes smaller, eventually to disappear altogether. The time scale for its disappearance is given by

$$t_{bh} = G^2 M^3 / \hbar c^4. \tag{10.10}$$

For a black hole of one solar mass, $t_{bh} = 10^{65}$ years.

A black hole radiates as if it were a black body with a temperature which is inversely proportional to its mass. Such a black body spectrum existed, as we have seen earlier, in the radiation in the early stages of the universe; it is describable in terms of a single temperature. The temperature of a black hole is of the order of $10^{26}/M$ K where M is the mass of the black hole in grams. For a supergalactic black hole this amounts to about 10^{-18} K. If the temperature of the cosmic background radiation is higher than this, the black hole will absorb more energy than it radiates. But as the universe expands, the temperature of the background radiation, which is proportional to $(R(t))^{-1}$, decreases. In the Einstein–de Sitter universe a temperature of 10^{-20} K would be reached in 10^{40} years, whereas in the dust universe with $k = -1$ (where R is asymptotically proportional to t) this temperature would be reached in 10^{30} years. For models with a positive cosmological constant this temperature would be reached earlier, since for these models R behaves exponentially (asymptotically) with time. Thus by the time galactic and supergalactic black holes are formed, or some time afterwards, the temperature of the black hole will exceed that of the background radiation and they will begin to radiate more than they absorb.

From (10.10) we see that a galactic black hole will last for about 10^{90} years while a supergalactic black hole will evaporate completely in about 10^{100} years. Thus after 10^{100} years or so black holes of all sizes will have disappeared, that is, all galaxies as we know them today will have been completely dissolved and the universe will consist of stray neutron stars, black dwarfs and smaller planets and rocks that were ejected from the galaxies. There will be an ever-increasing amount of empty space in which there will be a minute amount of radiation with an ever-decreasing temperature.

10.5 Slow and subtle changes

Consider the long term behaviour of any piece of matter, such as a rock or a planet, after it has cooled to zero temperature. Its atoms are frozen into an apparently fixed arrangement by the forces of cohesion and chemical binding. But from time to time the atoms will move and rearrange themselves, crossing energy barriers by quantum mechanical tunnelling. Even the most rigid materials will change their shapes and chemical structure on a time scale of 10^{65} years or so, and behave like liquids, flowing into spherical shape under the influence of gravity.

Any piece of ordinary matter is radioactive because it can release energy by nuclear fusion or fission reactions which take place by quantum tunnelling. All pieces of matter other than neutron stars must decay ultimately to iron, which has the most stable nucleus. The life-time for decay is given approximately by the Gamow formula $\exp[Z(M/m)^{1/2}]$, where Z is the nuclear charge, M the nuclear mass and m the electron mass. To get the actual life-time one has to multiply this pure number by some typical nuclear time scale, say 10^{-21} s. This gives a life-time of from 10^{500} to 10^{1500} years. On this time scale ordinary matter is radioactive and is constantly generating nuclear energy.

What will eventually happen to black dwarfs and neutron stars? If a black dwarf is compressed from outside by some external agent, it will collapse to a neutron star. In the near emptiness of the future universe there will be no external agent to compress it. However, the 'compression' can occur spontaneously by quantum tunnelling. The time scale can be calculated by another form of the Gamow formula, and is given as $10^{10^{76}}$ years (Dyson 1979). In a similar period, a neutron star will collapse into a black hole by quantum tunnelling and eventually evaporate by the Hawking process. Thus ultimately all black dwarfs and neutron stars will also disappear.

The decay of black dwarfs and neutron stars (indeed, of smaller pieces of matter) may occur earlier than $10^{10^{76}}$ years if black holes of smaller than stellar size are possible. Let M_B be the minimum size of a black hole, that is, suppose it is not, in principle, possible for a black hole to exist with mass less than M_B. Then the following alternatives arise:

(a) $M_B = 0$. In this case all matter is unstable with a comparatively short life-time.

(b) M_B is equal to the Planck mass: $M_B = M_{PL} = (hc/G)^{1/2} = 2 \times 10^{-5}$ g. This value of M_B is suggested by Hawking's theory, according to which every black hole loses mass until it reaches a mass of M_{PL}, at which point it disappears in a burst of radiation. In this case the life-time for all matter with mass greater than M_{PL} is $10^{10^{26}}$ years, while smaller pieces are absolutely stable.

(c) M_B is equal to the quantum mass $M_B = M_Q = hc/Gm_p = 3 \times 10^{14}$ g, where m_p is the proton mass. M_Q is the mass of the smallest black hole for which a classical description is possible (Harrison, Thorne, Wakano and Wheeler, 1965). In this case the life-time for a mass greater than M_Q is $10^{10^{52}}$ years.

(d) M_B is the Chandrasekhar mass $M_{ch} \simeq 4 \times 10^{33}$ g. In this case the life-time for a mass greater than M_{ch} is $10^{10^{76}}$ years, as mentioned earlier.

The long term future of matter in the universe depends crucially on which alternative is correct. Dyson (1979) favours (b). In the analysis so far we are assuming that the 'stable' elementary particles such as electrons and protons are, in fact, stable. This may not be the case over the periods which we have been discussing.

Barrow and Tipler (1978) show, under certain assumptions, that the universe will become increasingly irregular and unstable against the development of vorticity. This conclusion, however, is based on the assumption that the universe will consist of pure radiation in the long run, with all matter decaying. The matter density of stable matter varies as R^{-3} while that of radiation varies as R^{-4}. Thus radiation will dominate only if all matter decays. It is not clear how far this assumption is justified. Page and McKee (1981) find that a substantial proportion of the electrons and positrons (the latter arising from the decay of protons) will never annihilate.

The concept of the passage of time loses some of its meaning when applied to the final stages of the universe. Time is measured against some constantly changing phenomena. The only way in which the passage of time will manifest itself finally will be, presumably, the density and temperature of the background radiation, which will approach zero but never quite reach it.

The long term future of life and civilization has been discussed by Dyson (1979) (see also Islam (1979a,b, 1983a)).

10.6   A collapsing universe

The long term future of the universe is very different if the universe stops expanding and starts to contract. The life-time for a closed universe depends on the present average density of the universe.

Suppose the present density of the universe is twice the critical density. The universe will then expand to about twice its present size and start to contract. The total duration of the universe will be about 10^{11} years. The cosmic background radiation will go down to about 1.4 K and start to rise thereafter. The turning point will come in a few tens of billions of years – there will not be much change in the universe during this time. After the turning point, all the major changes that took place in the universe since the big bang will be reversed. In a few tens of billions of years, the cosmic background temperature will rise to 300 K, and the sky will be as warm all the time as it is during the day at present. After a few million years, galaxies will mingle with each other and stars will begin to collide with each other at frequent intervals. But before they get disrupted by such collisions, they will, in fact, dissolve because of the intensity of the background radiation (Rees, 1969), which will eventually knock out all electrons from atoms and finally neutrons and protons from nuclei. Ultimately, there will be a universal collapse of all matter and radiation into a compact space of infinite or near infinite density. It is not clear what will happen after such a collapse. Indeed, it is not clear if it is meaningful to talk about 'after' the final collapse, just as it is unclear whether it is meaningful to ask what happened 'before' the big bang.

In the steady state model proposed by Bondi and Gold (1948) and by Hoyle (1948) mentioned earlier, it is, in principle, possible for the universe to stay the same into the indefinite future. But as we have seen such a model is observationally untenable. It is also not clear in what way the above scenario is affected by the inflationary models, in which it appears possible to have different universes.

11

Some recent developments

11.1 Introduction

In this chapter we give a brief review of some recent papers which have appeared in connection with some of the topics dealt with in the earlier chapters. Here and there we also extend some of the earlier discussions. This review is by no means exhaustive but the reader can probably acquaint himself with a wide spectrum of recent papers on the topic by looking at the references contained in these papers, that have appeared in 1989 and 1990. Also, it is very difficult to assess the validity of the claims made in these papers in such a short time. We have mostly simply stated the results claimed; it is up to the reader to judge for himself by following up the reference concerned. Sometimes assessments of commentators have been given, such as those that appear in *Nature*.

11.2 Cosmic background radiation

One of the important observations relevant to cosmology in recent times was that carried out by the Cosmic Background Explorer (COBE) satellite (Lindley (1990); see also Carr (1988), Hogan (1990)). This satellite carried an instrument which was especially designed to measure the departure in the cosmic background radiation from a smooth 'reference' black body. As indicated earlier, any deviation from a smooth background, that is, any 'graininess' that is found, and its magnitude, can give useful information about primordial galaxy formation or other similar characteristics of the early universe. The range of wavelengths over which measurements were taken by the satellite was from 100 μm to 1 cm. It was found that departures from a black-body spectrum, if any, are less than 1%. The most recent observations by COBE, the results of which were presented at the April 1992 meeting of the American Physical Society in Washington DC (see *News and Views, Nature, Lond.* **356**, 741 (30 April 1992)), reveal slight departures from uniformity, the variation in temperature ΔT being given by $\Delta T/T = (5 \pm 1.5) \times 10^{-6}$. over angles up to 90°. This is an extremely important observation which is likely to have a significant effect on theories of galaxy formation.

11.3   Quasar astronomy

A significant advance in quasar astronomy (see Section 4.3) has been the observation of the optical spectra of the quasar Q1158 + 4635 (red-shift $z = 4.73$) and ten other quasars, with red-shifts $z > 4$ carried out by Schneider, Schmidt and Gunn (1989). Detailed statistical analysis remains to be done; these analyses are likely to provide clues to the physical conditions obtaining in the intergalactic medium in the very early evolution of the universe. An analysis of the fine structure in the absorption spectrum of a strong distant source such as a quasar can give useful information on types and concentrations of the intervening mass. This could possibly provide some clue to the problem of 'missing' or 'dark' matter.

11.4   Galactic distribution

Broadhurst *et al* (1990) (see also David (1990)) have studied large-scale distribution of galaxies at the galactic poles, both north and south. They find indications that galaxies are not distributed randomly but are clustered on scales of $5h^{-1}$ Mpc, where h is a constant denoting the uncertainty in the value of H_0; $H_0 = 100h$ km s^{-1} Mpc^{-1}, with a likely value in $0.5 \leqslant h \leqslant 1$. For this survey, data are taken from four different surveys at the north and south galactic poles. They find indications of periodic oscillations of density and evidence of structure at the largest scale studied by them. They emphasize the tentative nature of these observations, which need to be confirmed. If confirmed, these observations may have implications for theories of galaxy formation and for inflationary models.

11.5   New value of H_0

There has been a new estimate of the value of H_0 by Jacoby, Ciardullo and Ford (1990) (see also Fukugita and Hogan (1990)) which seems to be of considerable interest. As is clear from the earlier chapters, a correct observational determination of the value of H_0 is one of the most important problems in cosmology. As indicated earlier, the main difficulty here is to determine accurately the distance of galaxies which are relatively far; this is usually done by comparing their luminosity with that of standard candles such as Cepheid variables and Type Ia supernovae. The former exist only for nearby galaxies, while the latter are rare events. Jacoby *et al.* have been able to determine the distance to several galaxies in the Virgo cluster more accurately than before with the use of another type of standard candle, namely, planetary nebulae. The latter are clouds of radiating gas to which a star usually transforms towards the end of its life, when its hydrogen fuel is exhausted and it is burning only helium. The interesting thing is that there seems to be a maximum intrinsic brightness associated with planetary nebulae, the theoretical reason for which is not entirely clear; this could be to do with the maximum mass of the core of a star nearing its end – one which does not become a neutron star or a black hole – the so-called Chandrasekhar mass (around 1.4 solar masses) (see Section 10.2). Another advantage of the technique used by Jacoby *et al* seems to be that planetary nebulae seem to emit most of their energy in a narrow spectral band. This results in ease of detection and necessity of observing at a single epoch, unlike Cepheids. Hitherto the value of H_0 has been uncertain by a factor of about 2. Jacobi *et al.* claim to have calculated H_0 to within 15% in the range 75–100 km s^{-1} Mpc^{-1}, which is in the higher range of the previous uncertainty of 50–100 km s^{-1} Mpc^{-1}. This would have serious implications for cosmology, if confirmed. For example, this would imply that the Universe is somewhat younger than previously believed. (See (3.4), (3.25) and Section 3.2).

11.6   Cosmic book of phenomena

Peebles and Silk (1990) have compiled an interesting 'Cosmic book of phenomena' comparing five general theories for the origin of galaxies and large-scale structure in the universe by studying how well these theories are able to explain 38 different observational phenomena. This follows their earlier 'book' (1988) which dealt exclusively with large-scale structure. As mentioned in Chapter 3, estimates of the value of Ω, the density parameter (see (3.9)) based on observations and on the dynamics of systems of galaxies, yield a value somewhat less than unity, around 0.1. Theorists prefer a value close to unity, for reasons given in Section 8.1 ((8.1a), (8.1b)). The two points of view here are therefore, roughly speaking, firstly that $\Omega \approx 0.1$ with the mass density consisting mainly of ordinary (baryonic) matter and secondly, that the universe is dominated by some exotic non-baryonic matter which interacts weakly so that it is not readily detected (dark matter). Peebles and Silk examine the following five general theories which purport to explain the above scenarios, by seeing how well they deal with 38 different observational constraints. The first is the cold dark matter (CDM) theory (Frenk *et al*, 1988, 1990) in which the universe is Einstein–de Sitter (see Section 3.2), dominated by matter with negligible initial pressure (cold matter) that interacts weakly, and galactic structure emerges through suitable primeval density fluctuations. The hot dark matter (HDM) model (Zel'dovich, Einasto and Shandarin, 1982) has particles of dark matter with primeval velocity typical of neutrinos of mass about 30 eV; the remnant neutrinos make $\Omega \approx 1$ (see Section 7.8). In the string theories (STR) structure is formed by seeds of primeval non-linear perturbations; we shall come back to these theories. Weinberg, Ostriker and Dekel (1989) attempt to explain origin of structure in what Peebles and Silk call the explosion (XPL) picture, in which locally inserted energy, which could be from early supernovae, creates ridges of baryons which subsequently disintegrate to form new star clusters. In the baryonic dark matter (BDM) theory, unlike in the CDM theory, most of the galaxy masses were assembled at red-shifts $z \geqslant 10$. Peebles and Silk define a 'quality rating' parameter r, as follows

$$r = \tfrac{1}{2}(1 + 2wp - w), \tag{11.1}$$

where p is the probability for the theory and w is the weight for the phenomenon that is being explained. The parameter r has the character of a probability. If the weight w for the phenomenon is high, $w \approx 1$, then the rating r is nearly the same as p, the probability that the theory explains the phenomenon. If the weight is very low, $w \approx 0$, then $r \approx 0.5$, independent of p. Other cases fall in between these extreme cases. Peebles and Silk combine the r_i for 38 phenomena and compute the product $\prod r_i$, which is then used to determine the overall rating. An improbable theory would have a significant number of small r_i, whereas a 'good' theory would have more r_i near unity. Peebles and Silk find no clear winners but the CDM and BDM theories seem to them to be slightly ahead of the rest. As examples, we consider two of the phenomena in the list and give the weights w and the ratings r. The first one is that the isotropy of the cosmic background radiation is given by $\delta T/T < 2 \times 10^{-5}$ at around 30 arcmin. The weight w and the rating r for the five theories CDM, HDM, STR, XPL and BDM are respectively 1.0, 0.95, 0.05, 0.70, 0.70, 0.70. Secondly, for the phenomenon that there are clusters of galaxies as massive as the Coma cluster at $z = 1$, these quantities have the values 0.8, 0.14, 0.42, 0.86, 0.86, 0.86 respectively. These two phenomena are taken at random from the list; there are 38 such phenomena in the list, as mentioned earlier.

11.7   A critique of the standard model

Arp, Burbidge, Hoyle, Narlikar and Wickramasinghe (1990) are very critical of the standard model as described in the previous chapters and as believed by a great majority of cosmologists. Arp *et al.* cite various pieces of evidence to support their contention that, 'perhaps, there never was a Big Bang'. They also claim that the large red-shifts discovered so far, or at least substantial portions thereof, are in fact a result of intrinsic properties of the sources so that they do not lie at large cosmological distances, but are much closer, at distances that would follow from the Hubble Law for red-shifts $z \leqslant 0.1$. One of the reasons for this view is the discovery by Arp of cases of galaxies of very different red-shifts which are found very close together on the photographic plate. Opponents of this view contend that these are purely chance alignments of galaxies which are in reality very far from each other. Arp *et al* are aware of this criticism but they insist that their findings are statistically significant. Arp *et al* discuss at length the various other reasons for their lack of belief in the standard model. For example, they claim that the cosmic background radiation is not a relic of the primordial Big Bang, but is a result of the thermalization (that is, attainment of black-body spectrum) of the radiation given off after galaxy formation, and they suggest mechanisms through which thermalization could have occurred. They admit that they have no clear alternative for the standard model, but they suggest that a variation of the Steady State model (see Section 8.3), which can be considered as one of the forms of the scale-invariant conformal theory of gravitation put forward by Hoyle and Narlikar (see, for e.g., Hoyle and Narlikar (1974)), fits the current observations, as interpreted by them better. Various points Arp *et al* discuss are of intrinsic interest, whether or not their overall view is correct. Although this is a minority and an unpopular view, we believe such criticism is healthy for the subject of cosmology, for no theory or model should turn into a set of dogmas (Oldershaw, 1990). The onus is on the adherents of the standard model to provide adequate answers to these criticisms. Presumably some adherents would claim that adequate answers have already been given, but one can expect more answers to appear in the near future.

11.8   Cosmic strings

As mentioned earlier (see Section 8.5), among the possible relics of the phase transition of the very early universe are cosmic strings, which can be considered as thin lines of concentrated energy. If cosmic strings exist, they could be important for the formation of galaxies and large-scale structure of the universe. The evidence for cosmic strings is hard to find; this could come, for example, from gravitational radiation, which is notoriously difficult to detect. To be important for galaxy formation the mass per unit length of the strings should be in the region of

10^{22} g cm^{-1}, which is roughly the magnitude predicted by GUT. Such densities would produce certain potentially detectable observational effects, such as double images of distant galaxies and quasars due to gravitational lensing, certain discontinuities in the microwave background radiation, in addition to effects on gravitational radiation mentioned. The theoretical discussion of cosmic strings is difficult and interesting; they form a tangled web permeating the entire universe, with closed loops or extending to infinity without ends. Their evolution is believed to be scale-invariant; statistically the network is the same at all times. This implies that at any time t, the distance between nearby long strings is of the order of the horizon $\sim ct$ and typical loop size is a certain fraction of this distance. Recent simulations (Bennett and Bouchet (1989); Allen and Shellard (1990); see also Vilenkin (1990)) show that long strings have a significant fine structure on a scale somewhat smaller than the horizon, unlike what was believed earlier. A major portion of the structure is in the sharp angles, 'kinks', at points where strings are reconnected. The typical loop size l is also smaller than expected: $l \ll ct$. The new findings have interesting consequences for galaxy formation. There are two ways cosmic strings are believed to assist galaxy formation: gravitational attraction of loops and formation of wakes behind fast moving long strings; these were thought to be comparable. The new studies indicate, since loop sizes may be much smaller, that the second process may dominate. Further studies are needed to clarify various aspects of this interesting point.

11.9 Topological structures

Turok (1989; see also Friedman and Morris (1990)) has considered topological structures in the very early universe. As has been noted earlier, the astonishing uniformity of the cosmic background radiation is difficult to reconcile with the clumping of matter into galaxies and clusters. In Chapter 7 we saw that the background radiation is composed of radiation that left matter about 100 000 years after the big bang. As this radiation is isotropic to 1 part in 10^4 or so, the density variation around the period the radiation left matter could not have been significantly more than this fraction. It is difficult to evolve galaxies with such small variations unless one has exotic forms of matter. (In fact Arp *et al.* (1990), quoted earlier, cite this as a reason why galaxies should have been formed before the background radiation, although it is not clear if they can explain the extraordinary smoothness of the radiation.) Turok (1989) suggests that topological structures related to strings could provide ingredients for the formation of galaxies. It was mentioned in Chapter 8 that the symmetry of the four forces, namely, gravitation, electromagnetic, weak and the strong forces, presumably was broken successively through phase transitions in the very early universe. In addition to the example of freezing water cited in Chapter 8, one can consider the breaking of symmetry when a ferromagnet is cooled below 1043 K; this results in alignment of the randomly oriented spins, which form distinct domains, as in Fig. 11.1. As mentioned in Chapter 8, a similar breaking of symmetry may have occurred in the very early universe, which may be considered as being due to the appearance of Higgs fields. A topological structure may be associated with a Higgs field, which can be understood by considering a vector field pervading the universe, represented by an arrow of unit length at every point of the universe. Two configurations have different topologies if they cannot be deformed into each other by continuous changes. For example, if we consider a one-dimensional 'universe' (e.g. a circle), then the two configurations (a) and (b) of Fig. 11.2 cannot be continuously deformed into each other. These two configurations have different 'winding

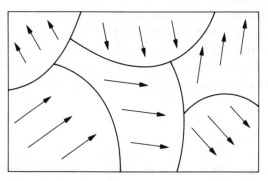

Fig. 11.1. When a ferromagnet is cooled below 1043 K, domains form with different magnetization.

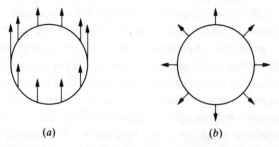

(a)   (b)

Fig. 11.2. The two configurations (a) and (b) cannot be deformed into one another through continuous transformations.

numbers'. (A typical element U of a group such as $SU(2)$ can be regarded as a mapping from S^3, the three-dimensional surface of a sphere in four-dimensional Euclidean space, onto the group manifold of $SU(2)$, that is, the space of parameters characterizing the group $SU(2)$, which is topologically the same as S^3. The *winding number* of a particular class of mappings is the number of times the spatial S^3 is covered by the group manifold S^3. Gauge transformations belonging to a group G, which can be any $SU(N)$, can be split up into homotopy classes, each of which is characterized by a distinct winding number). Essentially what may happen is that as a universe with a certain topological structure evolves, because this structure is preserved (cannot be made to 'go away') one may eventually get small regions of high energy density (called 'knots'), which may provide seeds for galaxy formation.

11.10   Extended inflation

In Chapter 8 we saw that one of the properties of the observed universe the inflationary models attempt to explain is the fact that Ω, the density parameter (see (8.1a), (8.1b)) is so close to unity. As mentioned earlier, the essential idea is that the universe spends a very short period in its very early history in a supercooled state, when a large constant and positive vacuum energy dominates its density of energy. The subsequent exponential expansion causes Ω to evolve towards unity. Also, inflation expands a causally connected region that is small into one that is much larger than the observable universe, thus solving the 'horizon' problem. In the 'old inflation' of Guth, there were 'bubbles' of the true vacuum in the supercooled state which could not merge and complete the phase transition. In the 'new inflation' this problem could perhaps be solved, but this required such 'fine tuning' of the parameters that it was not clear that such fine tuning could be achieved. Steinhardt (1990), proposes a model that he calls 'extended inflation' (see also Lindley 1990a), which, it is claimed, does not have the defects of earlier models in that there exist ranges of parameters which allow a set of initial conditions that lead to $\Omega \leqslant 0.5$, so that consistency with observation is obtained. Like in 'old inflation', in 'extended inflation' the barrier between the false and true vacuum is finite, but the new feature here is that the strength of gravitation varies with time, and this variation is related in a certain sense to the expansion of the universe. Steinhardt also shows that in the earlier 'new inflation' the fine tuning looked for could not have been achieved.

11.11   New cosmological solution

A new class of inhomogeneous cosmological solutions has been found by Senovilla (1990) which does not seem to possess any singularities in the past, with the curvature and matter invariants regular and smooth everywhere. The source is a perfect fluid with equation of state $\varepsilon = 3p$. The metric is as follows (with signature $+2$):

$$ds^2 = e^{2f}(-dt^2 + dx^2) + K(q\,dy^2 + q^{-1}\,dz^2), \tag{11.2}$$

where the functions f, K and q depend on t and x only and are given explicitly as follows:

$$e^f = [A\cosh(at) + B\sinh(at)]^2 \cosh(3ax),$$

$$K = [A\cosh(at) + B\sinh(at)]^2 \sinh(3ax)[\cosh(3ax)]^{-3/2}, \tag{11.3}$$

$$q = [A\cosh(at) + B\sinh(at)]^2 \sinh(3ax),$$

where a, A, B are arbitrary constants. The pressure and energy density are given as follows:

$$p = \tfrac{1}{3}\varepsilon = 5\chi^{-1}a^2[A\cosh(at) + B\sinh(at)]^{-4}[\cosh(3ax)]^{-4}, \tag{11.4}$$

where χ is the gravitational constant in suitable units. It is not clear how the absence of singularities can be reconciled with the singularity theorems (see Section 6.5). It is possible that this solution is not geodesically complete; the author says that this is under investigation.

11.12   More inflationary solutions

Ellis and Madsen (1991) find a number of exact cosmological solutions with a scalar field and non-interacting radiation, which could provide some new inflationary models (see Section 8.3). They give a method of 'generating' a class of solutions, following an old idea due to Synge (1955). We give two examples, in which the radiation density is set equal to zero, so that one has a pure scalar field (with $k = 1$ in (2.68)):

$$R(t) = A\exp(wt), \quad A, w \quad \text{constant} > 0, \tag{11.5a}$$

$$\phi(t) = \phi_0 \pm (B/w)\,e^{-wt}, \quad \phi_0 = \text{constant}, \quad B^2 = (4A^2\pi G)^{-1}, \tag{11.5b}$$

$$V(\phi) = (3w^2/8\pi h) + w^2(\phi - \phi_0)^2. \tag{11.5c}$$

This solution gives the usual de Sitter exponential expansion, without a singularity in the finite past, unlike the following solution, which has a singularity in the finite past

$$R(t) = A \sinh(wt), \quad A, w \quad \text{constant} > 0, \tag{11.6a}$$

$$\phi(t) = \phi_0 \pm (B/w) \log\left(\frac{e^{wt} - 1}{e^{wt} + 1}\right), \quad \phi_0 = \text{constant},$$

$$\tag{11.6b}$$

$$B^2 = \frac{1}{4\pi G}\left(w^2 + \frac{k}{A^2}\right) \geqslant 0,$$

$$V(\phi) = \frac{3w^2}{8\pi G} + B^2\left\{\sinh\left[\frac{2w}{B}(\phi - \phi_0)\right]\right\}^2. \tag{11.6c}$$

In (11.6b) the inequality is true for $k = 0, 1$ and can always be satisfied for $k = -1$. Ellis and Madsen discuss various properties of these solutions in the context of inflationary models. An interesting aspect of these solutions is that they allow a wide variety of behaviour for the density parameter Ω.

A generalization has recently been found by the author of the exact solution for an inflationary cosmology presented in Section 8.4, both for a sixth degree potential with a cubic term (which is absent in the solution given here; see (8.17)) and for potentials of higher degrees. It is likely that physically the new solutions will be more interesting; they involve a larger set of parameters which could be adjusted to give more realistic potentials. It is hoped to present this work in a future paper.

A number of interesting power law and exponential inflationary solutions, including ones which have intermediate expansion rates, have been considered by Barrow (1990), Barrow and Maeda (1990) and Barrow and Saich (1990).

11.13   Quantum cosmology

In Chapter 9 on quantum cosmology it was stated that the expression (9.37) for the amplitude has hidden in it many complexities, one of these being similar to that encountered in Yang–Mills theories which was dealt with by Faddeev and Popov (1967). In fact because of the indefinite metric and the nature of the space of geometries over which the path integral is taken, other complications arise of a different nature from that encountered in Yang–Mills theories. A satisfactory and precise formulation and definition of (9.37) (see also (9.55), (9.56)) still remains an important problem in quantum cosmology (see Halliwell and Hartle (1990), Halliwell and Louko (1990a,b)). An important aspect of the problem of quantum cosmology is that of 'decoherence', that is, the nature of the interference between different histories of the universe and the manner in which these effects eventually disappear to leave the universe to evolve classically subsequently (Gell-Mann and Hartle (1990); see also Calzetta and Mazzitelli (1990)).

11.14   Neutrino types

As discussed in Section 7.8, the number of types of neutrino is of cosmological importance. Among relevant points to emerge at the 14th International Conference on Neutrino Physics and Astrophysics at CERN in 1990 was that there are three neutrino types unless the mass of the fourth one exceeds 45 GeV; the relic abundance of such a heavy neutrino is not sufficient to contribute to dark matter (Griest and Silk, 1990; Salati, 1990). These results come from LEP, the Large Electron Positron collider at CERN.

References

Adams, E. N. 1988, *Phys. Rev.* **D37**, 2047.
Adams, F. C., Freese, K. and Widrow, L. M. 1990, *Phys. Rev.* **D41**, 347.
Albrecht, A. and Brandenberger, R. H. 1985, *Phys. Rev.* **D31**, 1225.
Albrecht, A. and Steinhardt, P. 1982, *Phys. Rev. Lett.* **48**, 1220.
Allen, B. and Shellard, E. P. S. 1990, *Phys. Rev. Lett.* **64**, 119.
Arp, H. C. 1967, *Ap. J.* **148**, 321.
Arp, H. C. 1980, in *Ninth Texas Symposium on Relativistic Astrophysics*, eds. J. Ehlers, J. Perry and M. Walker: New York Academy of Sciences, p. 94.
Arp, H. C., Burbidge, G., Hoyle, F., Narlikar, J. V. and Wikramasinghe, N. C. 1990 *Nature* **346**, 807.
Baade, W. 1952, *Trans. IAU* **8**, 397.
Bahcall, J. N. and Haxton, W. C. 1989, *Phys. Rev.* **D40**, 931.
Bahcall, J., Huebner, W., Magee, N., Mertz, A. and Ulrich, R. 1973, *Ap. J.* **184**, 1.
Bahcall, J. *et al.* 1980, *Phys. Rev. Lett.* **45**, 945.
Banks, T. 1988, *Nucl. Phys.* **B309**, 493.
Barrow, J. D. 1987, *Phys. Lett.* **B187**, 12.
Barrow, J. D. 1990, *Phys. Lett.* **235**, 40.
Barrow, J. D. and Maeda, K. 1990, *Nucl. Phys.* **B341**, 294.
Barrow, J. D. and Saich, P. 1990, *Phys. Lett.* **249**, 406.
Barrow, J. D. and Tipler, F. J. 1978, *Nature* **276**, 453.
Baum, W. A. 1957, *Ap. J.* **62**, 6.
Beer, R. and Taylor, F. W. 1973, *Ap. J.* **179**, 309.
Belinskii, V. A. and Khalatnikov, I. M. 1969, *Soviet Phys. JEPT* **29**, 911.
Belinskii, V. A., Khalatnikov, I. M. and Lifshitz, E. M. 1970, *Adv. Phys*, **19**, 525; 1970, *Soviet Phys. JETP* 33, 1061.
Bennett, D. P. and Bouchet, F. R. 1989, *Phys. Rev. Lett.* **63**, 2776.
Black, D. C. 1971, *Nature Phys. Sci.* **234**, 148.
Black, D. C. 1972, *Geochim. Cosmochim. Acta* **36**, 347.
Boato, G. 1954, *Geochim. Cosmochim. Acta* **6**, 209.
Bogoliubov, N. N. and Shirkov, D. V. 1983, *Quantum Fields*, Benjamin-Cummings Publishing Company, Inc., Reading, Massachusetts.
Bondi, H. 1961, *Cosmology*, Cambridge University Press, Cambridge, England.
Bondi, H. and Gold, T. 1948, *Mon. Not. Roy. Astr. Soc.* **108**, 252.
Bose, S. K. 1980, *An Introduction to General Relativity*. Wiley Eastern, New Delhi, India.

Brandenberger, R. H. 1985, *Rev. Mod. Phys.* **57**, 1.

Brandenberger, R. H. 1987, *Int. J. Mod. Phys.* **A2**, 77.

Brecher, K. and Silk, J. 1969, *Ap. J.* **158**, 91.

Broadhurst, T. J., Ellis, R. S., Koo, D. C. and Szalay, A. S. 1990, *Nature* **343**, 726.

Burbidge, G. R. 1981 in *Tenth Texas Symposium on Relativistic Astrophysics*, eds. R. Ramaty and F. C. Jones, New York Academy of Sciences, p. 123.

Burbidge, G. R. and Burbidge, M. 1967, *Quasi-Stellar Objects*, W. H. Freeman & Co., San Francisco.

Burbidge, G. R., Burbidge, M., Fowler, W. A. and Hoyle, F. 1957, *Rev. Mod. Phys.* **29**, 547.

Calzetta, E. and Mazzitelli, F. D. 1990, Preprint GTCRG-90-3, Grupo de Teorias Cuanti Relativistas y Gravitacion, University of Buenos Aires.

Carr, B. 1988, *Nature* **334**, 650.

Cesarsky, D. A., Moffet, A. T. and Pasachoff, J. M. 1973, *Ap. J. Lett.* **180**, L1.

Chandrasekhar, S. 1935, *Mon. Not. Roy. Astr. Soc.* **95**, 207.

Chandrasekhar, S. 1939, *An Introduction to the Study of Stellar Structure*, Dover Publications, New York.

Chandrasekhar, S. 1960, *Principles of Stellar Dynamics*, Dover Publications, New York.

Chernin, A. D. 1965, *Astr. Zh.* **42**, 1124.

Chernin, A. D. 1968, *Nature, Lond.* **220**, 250.

Chew, G. F. 1962, *S-Matrix Theory of Strong Interactions*, Benjamin, New York.

Chincarini, G. and Rood, H. J. 1976, *Ap. J.* **206**, 30.

Chincarini, G. L., Giovanelli, R. and Haynes, M. P. 1983, *Ap. J.* **269**, 13.

Coleman, S. 1988, *Nucl. Phys.* **B310**, 643.

Coleman, S. and Weinberg, E. 1973, *Phys. Rev.* **D7**, 788.

Conrath, B., Gautier, D. and Hornstein, J. 1982, *Saturn Meeting*, Tucson.

Crane, P. and Saslaw, W. C. 1986, *Ap. J.* **301**, 1.

Davies, P. C. W. 1973, *Mon. Not. Roy. Astr. Soc.* **161**, 1.

Davies, R. L. *et al.* 1987, *Ap. J.* **313**, 42, 15-D7; L-37, 27–49; **318**, 944, 91-C9.

Davis, M. 1990, *Nature, Lond.* **343**, 699.

de Vaucouleurs, G. 1977, *Nature, Lond.* **266**, 125.

Djorgovski, S. and Spinrad, H. 1981, *Ap. J.* **251**, 417.

Dyson, F. J. 1979, *Rev. Mod. Phys.* **51**, 447.

Eardley, D., Liang, E. and Sachs, R. 1972, *J. Math. Phys.* **13**, 99.

Eddington, P. 1930, *Mon. Not. Roy. Astr. Soc.* **90**, 668.

Einstein, A. 1950, *The Meaning of Relativity*, 3rd edn., Princeton University Press, Princeton, N.J.

Eisenhart, L. P. 1926, *Riemannian Geometry*, Princeton University Press, Princeton, N.J.

Ellis, G. F. R. and Madsen, M. S. 1991, *Class. Quant. Grav.* **8**, 667.

Faddeev, L. D. and Popov, V. N. 1967, *Phys. Lett.* **25B**, 29.

Feynman, R. P. 1948, *Rev. Mod. Phys.* **20**, 367.

Feynman, R. P. and Hibbs, A. R. 1965, *Quantum Mechanics and Path Integrals*, McGraw-Hill Book Company, New York.

Frenk, C. S., White, S. D. M., Davis, M. and Efstathiou, G. 1988, *Nature* **327**, 507.

Frenk, C. S., White, S. D. M., Efstathiou, G. and Davis, M. 1990, *Nature* **351**, 10.

Freund, P. G. O. 1986, *Introduction to Supersymmetry*, Cambridge University Press, Cambridge, England.

Friedman, I., Redfield, A. C., Schoen, B. and Harris, J. 1964, *Rev. Geophys.* **2**, 177.
Friedman, J. L. and Morris, M. S. 1990, *Nature* **343**, 409.
Fukugita, M. and Hogan, C. J. 1990, *Nature* **347**, 120.
Futamase, T. and Maeda, K. 1989, *Phys. Rev.* **D39**, 399.
Futamase, T., Rothman, T. and Matzner, R. 1989, *Phys. Rev.* **D39**, 405.
Gamow, G., 1948, *Nature* **162**, 680.
Gautier, D. *et al.* 1981, *J. Geophys. Res.* **86**, 8713.
Gautier, D. and Owen, T. 1983, *Nature, Lond.* **302**, 215.
Geiss, J. and Reeves, H. 1972, *Astron. and Astrophys.* **18**, 126.
Gell-Mann, M. and Hartle, J. B. 1990, in *Complexity, Entropy and the Physics of Information*, Santa Fe Institute for Studies in the Science of Complexity, Vol. IX, edited by W. H. Zurek, Addison Wesley.
Geroch, R. P. 1967, Singularities in the Spacetime of General Relativity: Their Definition, Existence and Local Characterization, PhD thesis, Princeton University.
Gödel, K. 1949, *Rev. Mod. Phys.* **21**, 447.
Goode, S. W. and Wainwright, J. 1982, *Mon. Not. Roy. Astr. Soc.* **198**, 83.
Gregory, S. A. and Thompson, L. A. 1982, *Ap. J.* **286**, 422.
Gregory, S. A., Thompson, L. A. and Tifft, W. G. 1981, *Ap. J.* **243**, 411.
Grevesse, N. 1970, *Colloque de Liège*, **19**, 251.
Griest, K. and Silk, J. 1990, *Nature, Lond.* **343**, 222.
Gunn, J. E. 1978 in *Observational Cosmology*, eds. A. Maeder, L. Martinet and G. Tammann: Geneva Observatory, Geneva, Switzerland.
Gunn, J. E. and Oke, J. B. 1975, *Ap. J.* **195**, 255.
Gunn, J. E. and Tinsley, B. M. 1975, *Nature, Lond.* **257**, 454.
Guth, A. 1981, *Phys. Rev.* **D23**, 347.
Halliwell, J. J. and Hartle, J. B. 1990, *Phys. Rev.* **D41**, 1815.
Halliwell, J. J. and Louko, J. 1990a, *Phys. Rev.* **D39**, 2206.
Halliwell, J. J. and Louko, J. 1990b, *Phys. Rev.* **D40**, 1868.
Harrison, B. K., Thorne, K. S., Wakano, M. and Wheeler, J. A. 1965, *Gravitation Theory and Gravitational Collapse*, University of Chicago Press, Chicago, USA.
Hartle, J. B. 1984, *Phys. Rev.* **D29**, 2730.
Hartle, J. B. 1986, 'Prediction in Quantum Cosmology', Lectures delivered at the 1986 Cargèse NATO Advanced Summer Institute, 'Gravitation and Astrophysics'.
Hartle, J. B. and Hawking, S. W. 1983, *Phys. Rev.* **D28**, 2960.
Hawking, S. W. 1975, *Comm. Math. Phys.* **43**, 199.
Hawking, S. W. and Ellis, G. F. R. 1973, *The Large-Scale Structure of Space-Time*, Cambridge University Press, Cambridge, England.
Hawking, S. W. and Penrose, R. 1970, *Proc. Roy. Soc.* **A314**, 529.
Heasley, J. and Milkey, R. 1978, *Ap. J.* **221**, 677.
Hewish, A., Bell, S. J., Pilkington, J. D. H., Scott, P. F. and Collins, R. A. 1968, *Nature, Lond.* **217**, 709.
Hirata, K. *et al.* 1987, *Phys. Rev. Lett.* **58**, 1490, 1798 (E).
Hodges, H. M. 1989, *Phys. Rev.* **D40**, 1798.
Hogan, C. J. 1990, *Nature, Lond.* **344**, 107.
Hoyle, F. 1948, *Mon. Not. Roy. Astr. Soc.* **108**, 72.
Hoyle, F. 1959 in *Paris Symposium on Radio Astronomy*, ed. R. N. Bracewell, Stanford University Press, Stanford, USA, p. 529.

Hoyle, F. and Narlikar, J. V. 1974, *Action at a Distance in Physics and Cosmology*, Freeman, San Francisco.

Huang, K. and Weinberg, S. 1970, *Phys. Rev. Lett.* **25**, 895.

Hubble, E. P. 1929, *Proc. Nat. Acad. Sci. US* **15**, 169.

Hubble, E. P. 1934, *Ap. J.* **79**, 8.

Hubble, E. P. 1936, *Realm of the Nebulae*, Yale University Press, New Haven, Connecticut, USA.

Huchra, J., Davis, M., Latham, D. and Tonry, J. 1983, *Ap. J. Suppl.* **52**, 89.

Iben, I. Jr. 1969, *Ann. Phys.* **54**, 164.

Islam, J. N. 1977, *Quart. J. Roy. Astr. Soc.* **18**, 3.

Islam, J. N. 1979a, *Sky and Telescope* **57**, 13.

Islam, J. N. 1979b, *Vistas in Astronomy* **23**, 265.

Islam, J. N. 1983a, *The Ultimate Fate of the Universe*, Cambridge University Press, Cambridge, England.

Islam, J. N. 1983b, *Phys. Lett.* **97A**, 239.

Islam, J. N. 1984, *Endeavour, New Series* **8**, 32.

Islam, J. N. 1985, *Rotating Fields in General Relativity*, Cambridge University Press, Cambridge, England.

Islam, J. N. 1989, *Proc. Roy. Soc.* **A421**, 279.

Islam, J. N. 1990a, 'An Exact Cosmological Solution Connecting Radiation and Matter Dominated Eras', preprint.

Islam, J. N. 1990b, 'An Exact Inflationary Solution for a Sixth Degree Potential', preprint.

Islam, J. N. and Munshi, N. I. 1990, 'On a Limit to the Cosmological Constant', preprint.

Jacoby, G. H., Ciardullo, R. and Ford, H. C. 1990, *Ap. J.* **356**, 332.

Jefferts, K. B., Penzias, A. A. and Wilson, R. W. 1973, *Ap. J.* **179**, L57.

Joshi, P. S. and Narlikar, J. V. 1986, *Int. J. Mod. Phys.* **1**, 243.

Karachentsev, J. D. 1966, *Astrofizica* **2**, 81.

Kardashev, N. 1967, *Ap. J.* **150**, L135.

Kasner, E. 1921, *Am. J. Math.* **43**, 126, 130, 217.

Khalatnikov, I. M. and Lifshitz, E. M. 1970, *Phys. Rev. Lett.* **24**, 76.

Kirshner, R. P., Oemler, A., Schechter, P. L. and Schectman, S. A. 1981, *Ap. J. Lett.* **248**, L57.

Kirzhnits, D. A. and Linde, A. D. 1972, *Phys. Lett.* **42B**, 471.

Kraan-Korteweg, R. C., Sandage, A. and Tammann, G. A. 1984, *Ap. J.* **283**, 24.

Kraft, R. 1961, *Ap. J.* **134**, 616.

Kristian, J., Sandage, A. and Westphal, J. A. 1978, *Ap. J.* **221**, 383.

Kruskal, M. D. 1960, *Phys. Rev.* **119**, 1743.

Kunde, V. *et al.* 1982, *Ap. J.* **263**, 443.

Kung, J. H. and Brandenberger, R. 1989, *Phys. Rev.* **D40**, 2532.

La, D. and Steinhardt, P. J. 1989, *Phys. Rev. Lett.* **62**, 376.

La, D., Steinhardt, P. J. and Bertschinger, E. W. 1989, *Phys. Lett.* **B231**, 231.

Lambert, D. 1967, *Observatory* **960**, 199.

Landau, L. D. and Lifshitz, E. M. 1975, *The Classical Theory of Fields*, 4th English edn., Pergamon Press, Oxford.

Landsberg, P. T. and Park, D. 1975, *Proc. Roy. Soc. Lond.* **A346**, 485.

Lemaître, G. 1927, *Ann. Soc. Sci. Brux.* **A47**, 49.

Lemaître, G. 1931, *Mon. Not. Roy. Astron. Soc.* **91**, 483.

Lifshitz, E. M. and Khalatnikov, I. M. 1963, *Adv. Phys.* **12**, 185.

Lifshitz, E. M. and Khalatnikov, I. M. 1971, *Sov. Phys. JETP Lett.* **11**, 123.

Lilley, S. J. and Longair, M. S. 1984, *Mon. Not. Roy. Astr. Soc.* **211**, 833.

Lilley, S. J., Longair, M. S. and Allington-Smith, J. R. 1985, *Mon. Not. Roy. Astr. Soc.* **215**, 37.

Linde, A. D. 1982, *Phys. Lett.* **108B**, 389.

Lindley, D. 1990, *Nature, Lond.* **343**, 207.

Lindley, D. 1990a, *Nature, Lond.* **345**, 23.

Longair, M. S. 1978 in *Observational Cosmology*, eds. A. Maeder, L. Martinet and G. Tammann, Geneva Observatory, Geneva, Switzerland.

Longair, M. S. 1983, *Phys. Bull.* **34**, 106.

MacCallum, M. A. H. 1973, in *Cargese Lectures in Physics*, Vol. 6. ed. E. Schatzmann, Gordon and Breach, New York.

Marshak, R. E., Riazuddin and Ryan, C. P. 1969, *Theory of Weak Interactions in Particle Physics*, Wiley-Interscience, New York.

Mazenko, G., Unruh, W. and Wald, R. 1985, *Phys. Rev.* **D31**, 273.

Mazzitelli, I. 1979, *Astr. Astrophys.* **80**, 155.

McCrea, W. H. and Milne, E. A. 1934, *Q. J. Math.* **5**, 73, 76.

McIntosh, J. 1968, *Mon. Not. Roy. Astr. Soc.* **140**, 461.

Miller, J. C. and Pantano, O. 1989, *Phys. Rev.* **D40**, 1789.

Misner, C. W. 1969, *Phys. Rev. Lett.* **22**, 1071.

Misner, C. W., Thorne, K. S. and Wheeler, J. A. 1973, *Gravitation*. W. H. Freeman and Company, San Francisco.

Morris, M., Thorne, K. and Yurtsever, U. 1988, *Phys. Rev. Lett.* **61**, 1446.

Narlikar, J. V. 1979, *Mon. Not. Roy. Astr. Soc.* **183**, 159.

Narlikar, J. V. and Padmanabhan, T. 1983, *Phys. Rep.* **100**, 152.

Oke, J. B. and Sandage, A. 1968, *Ap. J.* **154**, 21.

Oldershaw, R. L. 1990, *Nature, Lond.* **346**, 800.

Oort, J. H. 1983, *Annual Rev. Astron. Astrophys.* **21**, 373.

Orton, G. and Ingersoll, A. 1980, *J. Geophys. Res.* **85**, 5871.

Ostriker, J. and Tremaine, S. 1975, *Ap. J. Lett.* **202**, L113.

Pacher, T., Stein-Schabes, J. A. and Turner, M. S. 1987, *Phys. Rev.* **D36**, 1603.

Page. D. N. 1987, *Phys. Rev.* **D36**, 1607.

Page, D. N. and McKee, M. R. 1981, *Nature, Lond.* **291**, 44.

Pagel, B. 1984, *Phil. Trans. R. Soc.* **310A**, 245.

Peebles, P. J. E. 1966, *Phys. Rev. Lett.* **16**, 410.

Peebles, P. J. E. 1971, *Physical Cosmology*, Princeton University Press, Princeton, N.J.

Peebles, P. J. E. and Silk J. 1988, *Nature, Lond.* **335**, 601.

Peebles, P. J. E. and Silk J. 1990, *Nature, Lond.* **346**, 233.

Penzias, A. A. and Wilson, R. W. 1965, *Ap. J.* **142**, 419.

Petrosian, V. and Salpeter, E. E. 1968, *Ap. J.* **151**, 411.

Petrosian, V., Salpeter, E. E. and Szekeres, P. 1967, *Ap. J.* **147**, 1222.

Press, W. H. and Spergel, D. N. 1989, *Physics Today* (March).

Raychaudhuri, A. K. 1955, *Phys. Rev.* **98**, 1123.

Raychaudhuri, A. K. 1958, *Proc. Phys. Soc. Lond.* **72**, 263.

Raychaudhuri, A. K. 1979, *Theoretical Cosmology*, Oxford University Press, Oxford, England.

Rees, M. J. 1969, *Observatory* **89**, 193.

Rees, M. J. 1987, in *Proceedings of the International Conference on Mathematical Physics*, ed. J. N. Islam, University of Chittagong, Bangladesh.

Rindler, W. 1956, *Mon. Not. Roy. Astr. Soc.* **116**, 6.

Rogerson, J. B. Jr. and York, D. G. 1973, *Ap. J. Lett.* **186**, L95.

Rossi, G. C. and Testa, M. 1984, *Phys. Rev.* **D29**, 2997.

Ryan, M. P. Jr. and Shepley, L. C. 1975, *Homogeneous Relativistic Cosmologies*, Princeton University Press, Princeton, N.J.

Salati, P. 1990, *Nature, Lond.* **346**, 221.

Sandage, A. 1968, *Observatory* **88**, 91.

Sandage, A. 1970, *Physics Today*, February.

Sandage, A. 1972a, *Ap. J.* **173**, 485.

Sandage, A. 1972b, *Ap. J.* **178**, 1.

Sandage, A. 1972c, *Quart. J. R. Astr. Soc.* **13**, 282.

Sandage, A. 1975a, in *Galaxies and the Universe*: Eds. A. Sandage, M. Sandage and J. Kristian, University of Chicago Press, Chicago, p. 761.

Sandage, A. 1975b, *Ap. J.* **202**, 563.

Sandage, A. 1982, *Ap. J.* **252**, 553.

Sandage, A. 1987, *Proceedings of the IAU Symposium, Beijing, China.*

Sandage, A. and Hardy, E. 1973, *Ap. J.* **183**, 743.

Sandage, A., Katem, B. and Sandage, M. 1981, *Ap. J. Suppl.* **46**, 41.

Sandage, A. and Tammann, G. A. 1968, *Ap. J.* **151**, 531.

Sandage, A. and Tammann, G. A. 1969, *Ap. J.* **157**, 683.

Sandage, A. and Tammann, G. A. 1975, *Ap. J.* **197**, 265.

Sandage, A. and Tammann, G. A. 1983 in *Large Scale Structure of the Universe, Cosmology and Fundamental Physics*, First ESO-CERN Conference, eds. G. Setti and L. van Hove, Garching, Geneva, p. 127.

Sandage, A. and Tammann, G. A. 1986 in *Inner Space Outer Space*, eds. E. W. Kolb, M. S. Turner, D. Lindley, K. Olive and D. Steckel: University of Chicago Press, Chicago, p. 41.

Sandage, A., Tammann, G. A. and Yahil, A. 1979, *Ap. J.* **232**, 352.

Saslaw, W. C. 1973, *Publ. Astr. Soc. Pacific* **85**, 5.

Saslaw, W. C., Valtonen, M. J. and Aarseth, S. J. 1974, *Ap. J.* **190**, 253.

Schmidt, B. G. 1973, *Commun. Math. Phys.* **29**, 49.

Schneider, D. P., Schmidt, M. and Gunn, J. E. 1989, *Ap. J.* **98**, 1507, 1951.

Schramm, D. N. 1982 paper at Royal Society (London) meeting 11–12 March.

Schramm, D. N. and Wagoner, 1974, *Physics Today*, December.

Schücking, E. and Heckmann, J. 1958, World Models in *Onzieme Conseil de Physique Solvay*, Editions Stoops, Brussels, pp. 149–58.

Schwarzschild, B. 1989, *Physics Today*, March.

Seldner, M., Siebers, B., Groth, E. J. and Peebles, P. J. E. 1977, *Ap. J.* **82**, 249.

Senovilla, J. M. M. 1990, *Phys. Rev. Lett.* **64**, 2219.

Shellard, E. P. S. and Brandenberger, R. H. 1988, *Phys. Rev.* **D38**, 3610.

Smoluchowski, R. 1967, *Nature, Lond.* **215**, 691.

Spaenhauer, A. M. 1978, *Astron. Astrophys.* **65**, 313.

Spinrad, H. 1986, *Pub. A.S.P.* **98**, 269.

Steinhardt, P. J. 1990, *Nature, Lond.* **345**, 47.

Synge, J. L. 1937, *Proc. Lond. Math. Soc.* **43**, 376.

Synge, J. L. 1955, *Relativity: the General Theory*, North-Holland: Amsterdam.

Szafron, D. A. 1977, *J. Math. Phys.* **18**, 1673.

Szafron, D. A. and Wainwright, J. 1977, *J. Math. Phys.* **18**, 1668.

Szekeres, P. 1975, *Commun. Math. Phys.* **41**, 55.

Tammann, G. A., Yahil, A. and Sandage, A. 1979, *Ap. J.* **234**, 775.

Tarenghi, M., Tifft, W. G., Chincarini, G., Rood, H. J. and Thompson, L. A. 1979, *Ap. J.* **234**, 793.

Tayler, R. J. 1983, *Europhysics News* **14**, 1.

Taylor, J. C. 1976, *Gauge Theories of Weak Interactions*, Cambridge University Press, Cambridge, England.

Thorne, K. S. 1967, *Ap. J.* **148**, 51.

Thorne, K. S. 1974, *Scientific American*, December.

Tinsley, B. M. 1977, *Physics Today*, June.

Tinsley, B. M. 1978, *Nature, Lond.* **273**, 208.

Trauger, J. T., Roesler, F. L., Carleton, N. P. and Traub, W. A. 1973, *Ap. J. Lett.* **184**, L137.

Turner, M. S. 1985 in *Proceedings of the Cargese School on Fundamental Physics and Cosmology*, eds. J. Audouze and J. Tran Thanh Van: Editions Frontièrs, Gif-Sur-Yvette).

Turok, N. 1989. *Phys. Rev. Lett.* **63**, 2625.

Ulrich, R. and Rood, R. 1973, *Nature Phys. Sci.* **241**, 111.

Van den Bergh, S. 1975, in *Galaxies and the Universe*, University of Chicago Press, Chicago.

Vilenkin, A. 1990, *Nature, Lond.* **343**, 591.

Wagoner, R. V. 1973, *Ap. J.* **179**, 343.

Wagoner, R. V., Fowler, W. A. and Hoyle, F. 1967, *Ap. J.* **148**, 3.

Wainwright, J. 1979, *J. Phys. A* **12**, 2015.

Wainwright, J. 1981, *J. Phys. A* **14**, 1131.

Wainwright, J. and Goode, S. W. 1980, *Phys. Rev.* **D22**, 1906.

Wainwright, J., Ince, W. C. W. and Marshman, B. H. 1979, *Gen. Rel. Grav.* **10**, 259.

Wainwright, J. and Marshman, B. J. 1979, *Phys. Lett.* **72A**, 275.

Weinberg, D. H., Ostriker, J. P. and Dekel, A. 1989, *Ap. J.* **336**, 9.

Weinberg, S. 1972, *Gravitation and Cosmology*, John Wiley and Sons, New York.

Weinberg, S. 1977, *The First Three Minutes*, André Deutsch, London; re-published in 1983 with an Afterword by Fontana Paperbacks.

Weinreb, S. 1962, *Nature, Lond.* **195**, 367.

Weiss, N. 1989, *Phys. Rev.* **D39**, 1517.

Weyl, H. 1923, *Phys. Z.* **24**, 230.

Wilson, R. W., Penzias, A. A., Jefferts, K. B. and Solomon, P. R. 1973, *Ap. J. Lett.* **179**, L107.

Zel'dovich, Ya. B. 1968, *Uspekhi* **95**, 209.

Zel'dovich, Ya. B., Einasto, J. and Shandarin, S. F. 1982, *Nature, Lond.* **300**, 407.

Index